Escaping the Energy Povert

Escaping the Energy Poverty Trap

When and How Governments Power the Lives of the Poor

Michaël Aklin, Patrick Bayer, S.P. Harish, and Johannes Urpelainen

The MIT Press
Cambridge, Massachusetts
London, England

This book was set in ITC Stone Sans Std and ITC Stone Serif Std by Toppan Best-set Premedia Limited. Printed and bound in the United States of America.

Library of Congress Cataloging-in-Publication Data

Names: Aklin, Michael, author. | Bayer, Patrick, 1983- author. | Harish, S. P., author.
Title: Escaping the energy poverty trap : when and how governments power the lives of the poor / Michael Aklin, Patrick Bayer, S. P. Harish, and Johannes Urpelainen.
Description: Cambridge, MA : MIT Press, [2018] | Includes bibliographical references and index.
Identifiers: LCCN 2018008756| ISBN 9780262038799 (hardcover : alk. paper) | ISBN 9780262535861 (pbk. : alk. paper)
Subjects: LCSH: Energy policy. | Energy security. | Poor--Government policy.
Classification: LCC HD9502.A2 U77 2018 | DDC 333.79--dc23 LC record available at https://lccn.loc.gov/2018008756

10 9 8 7 6 5 4 3 2 1

Contents

Preface

This book is the result of a project that began in New York in March 2012. The four of us met at a bar next to the New York University campus, where Michaël and Harish were graduate students at the time, to discuss new ways to study environmental politics. Our starting point was that a lot had been written about environmental issues in wealthy countries but we knew less about the sustainability challenge in emerging and developing countries.

This meeting led to a research program that took us to three continents. From Brazil to India to Tanzania, we tried to understand the nature and causes of unsustainable economic growth. In doing so, we soon learned how difficult it would be to disentangle environmental degradation from the broader challenge of economic development.

In this context, our work began to prioritize the problem of energy poverty. Billions of people lead lives without electricity at home or access to clean cooking fuels. These people's lack of access to modern energy often makes life cumbersome and unproductive. Moreover, to talk about sustainable development when billions have no access to affordable and abundant energy is, and will remain, a political nonstarter in the countries we studied.

Solving the problem of energy poverty is not enough for sustainable development, but it is necessary. As we wrote *Escaping the Energy Poverty Trap*, we began to understand how important government policies, and the messy politics surrounding them, would be for ending energy poverty. This book is our attempt to understand why governments sometimes succeed in improving energy access, why they sometimes fail, and how they could do better in the future.

As we have written this book, we have accumulated the usual authors' debts. We are immensely grateful to all our collaborators and partners across the world. Arunabha Ghosh, who leads the Council on Energy,

Environment and Water, a dynamic research institute located in New Delhi, India, invited Johannes to visit India in October 2012. That visit gave rise to our field research on rural energy access. Rustam Sengupta, who at that time led Boond Engineering & Development, a social enterprise focused on energy access in India, has collaborated with us extensively for the last several years. Beth Clevenger at the MIT Press, who had worked with Michaël and Johannes before, did a terrific job at guiding us through the publishing process. Douglas Barnes, Sunila Kale, and Brian Min gave excellent comments on different chapters of the manuscript. Most importantly, we are grateful to the people who generously gave their time to participate in our surveys and experiments in the field—they taught us most of what we know about energy poverty.

—The authors

Much of this book was written in fairly turbulent personal times: our son Emil was born in 2015, my wife Carolina and I were on the job market, and we moved from the US to Scotland. During these times of change, it was my wife's enduring support and love that made everything easy and joyful. I would also like to thank my now three-year-old son Emil for showing me every day that answering the seemingly easy questions is the most difficult part of our lives—both personally and academically. They are the loves of my life.

—Patrick

Academia is enriching in many ways, but it imposes unusual constraints. I am grateful to Sarah, my spouse, for her extraordinary support. To write a book like this requires the occasional midnight phone call while on vacation. Being supportive in these circumstances is a form of true love. Thank you.

—Michaël

1 Introduction

In today's industrialized world, almost everything we do requires huge amounts of energy. We use natural gas to heat our homes and cook our food. Electricity powers our fridges, freezers, mobile phones, tablets, computers, and televisions. Our malls, restaurants, offices, and streets are well lit by countless electric lights. Nuclear power stations, natural gas turbines, coal power plants, and huge dams churn out more power than ever before, and as we buy larger houses and use new electric appliances, we become more and more dependent on abundant and affordable energy. If we suddenly lost our ability to produce copious amounts of energy, our industrial economies would come to a standstill. Our survival, lifestyle, and identity depend on energy.

Elsewhere in the world, many lead lives impoverished by a lack of modern energy. More than 1 billion people do not have any electricity at home, and almost 3 billion people still use traditional fuels like biomass for cooking (IEA 2017). Much of the convenience and comfort that even the poorest people in industrialized countries take for granted is beyond the grasp of the billions who are among the world's energy poor. Energy poverty has tragic consequences for their lives. Without access to modern energy, the aspirations of the energy poor of the world are bound to be frustrated. Lack of electricity prevents the use of bright lighting for reading and studying (Khandker, Barnes, and Samad 2009). Without refrigerators, households cannot store perishable food or lifesaving medicine; absent electric power, small businesses are difficult to run (Cabraal, Barnes, and Agarwal 2005). Without a fan, oppressive heat in the tropics is a major threat to public health and productivity. Indoor air pollution from cooking with traditional stoves using firewood or other traditional biomass is a major health hazard for families (Ellegard 1996; Wickramasinghe 2011). Women and children spend hours upon hours to collect the same firewood that later spoils their lungs (Foell et al. 2011). If households had access to modern cooking fuels such as liquefied petroleum gas (LPG), these problems could be easily avoided.

How can the world's energy-poor gain access to abundant energy? Why have some countries, even among the least developed, left energy poverty behind? Why do others struggle to provide their people with energy? What can governments do to provide the poorest of the poor with access to modern energy? Unfortunately, the answers to these questions remain elusive. Scholars have identified energy resources, geography, and economic development as fundamental drivers of energy poverty. And yet these three factors can hardly explain patterns of energy poverty in the world. Against all odds, a gigantic country such as China made rapid progress in rural electrification during the first decade after the communist revolution. Vietnam's rural electrification program in the aftermath of total destruction from war is the envy of the world. In Brazil, the transformation from traditional biomass to LPG began almost a century ago. Yet even resource-rich countries such as Nigeria can struggle to provide energy to its citizens. Resources, geography, and economic development are not sufficient on their own to explain energy poverty.

Drawing on years of field research and data analysis, this book is our effort to solve the conundrum of energy poverty. Focusing on household electrification and access to modern cooking fuels, we show that a political economy approach to energy poverty can explain the varied outcomes across the developing world. Even after accounting for existing explanations, large differences across countries and over time remain, and these differences can be largely attributed to political factors. The three building blocks of our argument are government interest, institutional capacity, and local accountability.

Fundamentally, both political and economic factors determine the degree of a *government's interest* in eradicating energy poverty. Some governments expect economic gains from acting to erase energy poverty, while others do not. Nevertheless, governments of all stripes, both democratic and autocratic, are interested in the political benefits of eradicating energy poverty. In some circumstances, the energy-poor can shape the fates of political elites, be it through elections or protest, and these circumstances encourage progress toward eradicating energy poverty in a world of self-interested politicians.

Government interest is a necessary condition for successful policies to mitigate energy poverty, yet it is by no means sufficient. Among interested governments, two factors turn out to be decisive: *institutional capacity* and *local accountability*. Institutional capacity determines the ability of the government to enact and implement coherent, effective, and affordable policies that eradicate energy poverty over time. Local accountability ensures that national policies are properly implemented in energy-poor communities in terms of both information transmission and the incentives of local officials. In other words, local accountability ensures that policies reflect the specific needs and demands of the people.

The previously neglected political economy dimension is important for energy poverty from analytical and practical perspectives. It explains to a large extent the puzzling variation in energy poverty. Furthermore, because political economy focuses on government behavior and effective policy, the result implies much room for improvement in the future. Our findings provide ample reason for social scientists to focus on energy poverty, as social science can generate knowledge about strategies to induce governments to act effectively to end the tragedy of energy poverty. Furthermore, much in the domain of energy poverty will be of interest to the broad community of scholars working on political economy and development, such as the importance of distinguishing between the practical challenges of providing various types of public goods and services.

Importantly, this political economy approach can be used to understand the role of rapidly evolving technology in the eradication of energy poverty. Today, improvements in energy technology, such as ultraefficient LED lights and inexpensive solar panels, promise new opportunities for dealing with energy poverty. As we show, however, the ability of new technologies to eradicate energy poverty depends on government interest, institutional capacity, and local accountability. New technologies open new opportunities for both government-led and private initiatives, such as the creation of bottom-of-the-pyramid energy markets (Prahalad 2006; Shukla and Bairiganjan 2011), but their ability to make a difference depends on the same fundamental political economy factors that determine improvements in energy access more generally.

The remainder of this introduction has five goals, and each warrants a section of its own. The first goal is to describe variation in energy poverty, explain why we all should care, and motivate our political economy approach. The second is to summarize our argument, and the third is to describe our data and research methodology. Fourth, we discuss the implications of our findings and summarize the contributions of this book. As is often the case, the chapter ends with an outline of the book.

Energy Poverty and Human Development

Energy is a necessity for modern economic activity and lifestyle. Today, industrialized economies use vast amounts of energy to produce goods and services. We use electricity for lighting, to power appliances, and to run factories. For cooking and heating, we use both electricity and alternatives such as natural gas and LPG. We also use energy for transportation, and today's mechanized agriculture is dependent on large quantities of it. According to IEA statistics, in 2015 the average American used energy amounting to that released from burning 6,801 kilograms of oil.[1]

The majority of the world's population, however, has to survive with much less energy. The same IEA statistics tell us that in 2014, the average person consumed less than the equivalent of 1,920 kilograms of oil. In sub-Saharan Africa, the number was only 687 kilograms, and in South Sudan, where energy consumption per capita is the lowest in the world, this number was only 61 kilograms—or less than 1 kilogram for every 100 kilograms in the United States. The inequalities in energy consumption across countries are stark: some people have access to abundant energy, and others suffer from extreme deprivation.

If anything, other statistics tell an even clearer story. The IEA (2017) maintains a comprehensive database on household energy access in more than a hundred countries in the world. According to this database, 1.1 billion people in the world in 2016 remained without any access to electricity at home, and virtually all of these people were found in sub-Saharan Africa or the developing countries of Asia. Their children do not study at night using bright electric lights; they cannot charge their mobile phones at home or turn on the fan in the oppressive heat. Simple pleasures that we take for granted, such as watching television, are not available to them (Bernard 2010; Sagar 2005; World Bank 2008; Khandker, Barnes, and Sama 2013).

According to this 2017 IEA report, 2.8 billion people still lack access to modern cooking fuels and must rely on traditional biomass and other solid cooking fuels for their primary cooking needs. The large population of Asian developing countries leaves 1.65 billion (43 percent of the regional population) without access to modern cooking fuels; in sub-Saharan Africa, 780 million (80 percent of regional population) are in the same situation. Because most of this traditional biomass is firewood used for cooking, people living in these households spend a lot of time collecting their fuel (Arnold, Köhlin, and Persson 2006). Furthermore, indoor air pollution from firewood is a major cause of illness and death from respiratory diseases in the developing world (Kandpal, Maheshwari, and Kandpal 1995; Ellegard 1996; Mishra, Retherford, and Smith 1999; Goldemberg et al. 2004; Schei et al. 2004). As the director of the US National Institutes of Health, Francis S. Collins, said at the launch of the Global Alliance for Clean Cookstoves, "Indoor cookstoves are a serious health concern in developing countries. The toxic emissions from these cooking fires cause low birth weights, pneumonia in young children, and heart and lung problems in adults, and kill nearly 2 million people each year, mostly women and children."[2]

Our goal in this book is to understand and explain the phenomenon of energy poverty in rural households, for the evidence is clear: energy poverty is mostly an affliction outside urban areas. We also focus on two specific facets of energy poverty: (1) access to electricity and (2) access to modern cooking fuels and technology. These aspects of energy poverty make people poor and sick and warrant particular attention. While energy poverty cannot be reduced to these two dimensions—for instance, modern energy allows people to travel at unprecedented speed across long distances—we believe that electricity access and cooking are the two most pressing problems related to energy poverty, and thus our focus. Indeed, the UN Sustainable Development Goals include "affordable and clean energy."[3]

Some countries, such as Ghana and Vietnam, made early and rapid progress in the eradication of energy poverty despite their low levels of income and infrastructural development. Others, such as Nigeria and Bangladesh, have achieved much less despite similar circumstances and, if anything, more favorable conditions. How can we explain this variation? Why have some governments acted decisively and effectively to provide people with basic energy access and beyond, while others remain passive or are failing in their efforts?

Now is the time to answer these questions. The United Nations first declared the year 2012 the year of Sustainable Energy for All (SE4ALL) and then honored the years 2014 to 2024 with the same designation. For President Barack Obama, the Power Africa program was a foreign policy priority. From Pakistan and India to East Africa, governments continue to invest staggering amounts in grid extension and new power plants, while ambitious entrepreneurs are innovating with off-grid renewable power generation and technologically advanced cookstoves. Energy poverty remains a vexing problem, but there has never before been as much interest in the topic as there is today in international fora and among young entrepreneurs. The stakes for solving the problem are high too. For example, the latest IEA *World Energy Outlook* suggests that obtaining full energy access by 2030 would avoid about 1.8 million premature deaths from indoor air pollution alone (IEA 2017, 72).

The policy recommendations on ending energy poverty are decidedly controversial and embattled. Climate policy advocates and environmentalists are enthusiastic about the use of renewable resources, such as solar power, for rural electrification, but they worry about the negative effects of coal power plants on the global environment (Alstone, Gershenson, and Kammen 2015; ODI 2016). Many development specialists point to the tragic consequences of energy poverty for billions in the least developed countries (BTI 2014). These debates become particularly heated when concrete policies are at stake. From the World Bank to the US Overseas Private Investment Corporation, governments of industrializing countries have placed restrictions on investment into fossil fuels (Moss, Pielke, and Bazilian 2014). IEA's *World Energy Outlook* (2017) points in the right direction by developing a "Sustainable Development Scenario" that seeks to address decarbonization, energy access, and air pollution all in one.

Given this debate, a comprehensive analysis of the political economy of energy poverty is of much value. For researchers, the analysis of energy poverty is a new frontier that can shed light on deep questions about development more generally. For policymakers and practitioners, a thorough understanding of the political incentives and constraints that guide governments in energy-poor countries is absolutely critical for productive engagement. Programs such as SE4ALL are ultimately dependent on the participating governments, and these governments face political pressures as they make decisions about investments in energy access. Investments are key as an estimated US$52 billion would be needed annually to make energy for all a reality by 2030 (IEA 2017, 13).

The Political Economy of Energy Poverty

To explain patterns of energy poverty and solve these puzzles, we begin with existing explanations from the literature. Throughout, we focus on the contemporary problem of providing access to modern energy to a large number of poor people in developing countries. As fundamental factors, authors from different disciplines have emphasized the availability of domestic energy resources (Madureira 2008; Bhutto and Karim 2007), geography (Deichmann et al. 2011; Kemmler 2007; Onyeji, Bazilian, and Nussbaumer 2012), and the level of economic development (Foley 1992; Barnes 2011; Golumbeanu and Barnes 2013; Khandker, Barnes, and Samad 2012). While the availability of energy and favorable geographies makes the eradication of energy poverty easier, they cannot account for slow progress in a country such as Nigeria, which enjoys abundant energy wealth and has a very high population density. Economic development is a fundamental driver of energy access, yet countries such as China and Vietnam moved forward in leaps and bounds decades ago when they were still among the poorest in the world.

Others attribute variation in energy poverty to political institutions, such as the autocracy-democracy difference (Brown and Mobarak 2008) or emphasize the importance of effective policy (Haanyika 2006; Barnes 2007; Mawhood and Gross 2014; Nygaard 2010). But some autocracies, including China and Vietnam, have made great progress in eradicating energy poverty. So did Eastern European communist regimes under the yoke of the Soviet Union. What is more, effective policy is a tautological argument, as it fails to answer why all governments do not enact and implement effective policies.

There is no doubt that these factors go a long way toward explaining energy poverty. For example, while having coal or being a small country makes the problem easier to solve, such advantages do not automatically solve the problem nor does their absence mean that all hope is lost. The fundamentals of resources, wealth, and geography are neither necessary nor sufficient for mitigating energy poverty.

What is missing from the large body of literature on energy poverty is a systematic account of the political economy behind it. We argue that little progress is to be expected in the absence of decisive political action. By "political economy," we refer to the analytical tradition that emphasizes societal interests and their aggregation through political institutions into policy. In this line of reasoning, the focus is on government behavior. Governments are assumed to pay close attention to the prospect of political survival through the maximization of political support in the society. Consequently, governments enact and implement policies that cater to politically powerful groups. These policies may or may not be good for the energy-poor, but most of the time they are good for the government itself.

From a political economy perspective, the first factor that we consider is *government interest* in eradicating energy poverty, that is, the constellation of factors that encourage a government to invest resources into improving energy access and mitigating energy poverty. These factors can be economic or political. If measures that combat energy poverty also generate direct economic benefits, any government—no matter how self-interested—sees value in action. For example, rural electrification offers proceed in tandem with efforts to electrify irrigation pumps in the countryside for greater economic productivity. What is more, the energy-poor are sometimes in a position to influence the policymaking process because of their political power. For example, if the energy-poor live in areas that are pivotal in elections, the government has an incentive to combat their plight. We mostly focus on national governments, as key energy access policies are formulated at this level in most countries, but in some cases, we also consider the provincial level. In India, for example, the constitution gives states substantial powers over the distribution of electricity to users.

Nevertheless, government interest itself is not a guarantee of success. Historically, many governments have expressed genuine interest in mitigating energy poverty but failed to achieve their goals. Therefore, we must dig deeper into the political economy of energy poverty to understand why governments sometimes fail to realize their commendable goals. Without government interest, it is clear that a policy solution to the problem of energy poverty is not possible. Conditional on sufficient government interest, we must shift our attention to factors that determine success or failure in policy effort. Accordingly, the two other pieces of the puzzle that we highlight are *institutional capacity* and *local accountability*. Improvements in each factor are conducive to the mitigation of energy poverty.

By "institutional capacity," we refer to the government's access to an administrative apparatus that is capable of implementing policies in a competent and cost-effective manner (Bekker et al. 2008). Consider the following examples. Some governments have well-managed electric utilities and competent rural electrification agencies, whereas others must do with fiscally insolvent utilities and rural electrification agencies that do not have any actual capacity to implement policy across the country. Some governments have distribution networks for LPG to remote rural communities, while others cannot reach these communities at an acceptable cost. Whatever the government plans to do, institutional capacity determines the quality and cost of implementation.

Local accountability is equally important. Although governments can, in principle, enact and implement policies in a top-down manner, experience suggests that local participation is essential to success. While the notion of government interest, as defined in this book, focuses on higher levels of the political system, we identify a specific role for local politics and governance. Although local bodies rarely have influence over key energy access policies, they often play an important role in the implementation of higher-level policies on the ground. Thus, the local accountability of government officials is critical. By this, we refer to the extent that the community is able to voice its concerns and hold the local government accountable for their policies, successes, and failures (Paul 1992; Fox 2015). Even a government with a strong political or economic interest in removing energy poverty benefits from municipal or other local officials who can support the required public service delivery. Local accountability mechanisms improve information transmission from the grassroots level upward in the political machinery and ensure that the local government is rewarded for success and punished for failure. Therefore, local accountability mechanisms facilitate the eradication of energy poverty.

Without delving into any particular national circumstances, we can already make one prediction that turns out to be essential for understanding the mitigation of energy poverty: government policy is more effective at improving rural electrification than at providing modern cooking fuels to people who rely on traditional biomass. The reason is that rural electrification provides different governments with many opportunities to reap direct economic gains, whereas the provision of modern cooking fuels does not, despite clear environmental and public health benefits in the long run. This expectation is consistent with differential progress in rural electrification and the provision of clean cooking fuels. While the number of people with electricity access across the globe fell from 1.7 billion in 2000 by about 600,000 by 2016, the number of people lacking access to modern cooking fuels stagnated at around 2.8 billion over the same time period (IEA 2017, 11).

The theory is also relevant to understanding, predicting, and improving the role of rapidly evolving energy technology in the eradication of energy poverty. Given today's innovative dynamism in the energy sector, no theory of energy poverty can avoid commenting on technology. We do not expect technological progress, such as the reduced cost of off-grid solar systems, to fundamentally change the political economy logic we have outlined, but better technology does reduce the cost of government policy in all circumstances and therefore creates new opportunities for the eradication of energy poverty. Where the alignment of political economy factors is conducive to good policy, new energy technology can increase the pace of progress in the provision of universal energy access. But where governments lack the interest or ability to enact effective policies, new technology will not make a large difference in efforts to expand energy access.

Research Methods

Energy poverty is a relatively novel topic for political scientists. While there is a large body of literature by energy engineers and economists (Revelle 1976; Pachauri and Spreng 2004; Pachauri and Jiang 2008; Samad et al. 2013), political scientists have, with few exceptions (examples are Aklin et al. 2014b; Min and Golden 2014; Min 2015), not examined the political origins of the problem. Due to the limited availability of prior research, we experiment with our research design and methodological approaches. Given the complexity of the topic, we opt for the mixed-methods approach (Seawright and Gerring 2008). For this, we use a variety of qualitative methods, ranging from a detailed case study (Gerring 2004), to a subnational comparative analysis, to a series of structured and focused comparisons, in order to gain deeper insights into causal mechanisms and processes (George 1979; King, Keohane, and Verba 1994; Sekhon 2004).

Our first challenge is to provide a proper description of energy poverty across the world today. We conduct one major national case study—India— to gain a deep understanding of the political economy of energy poverty and eleven less comprehensive case studies for the purposes of comparative hypothesis testing. Focusing on India is ideal for a comprehensive case study for four reasons. First, it has a long history of efforts to reduce energy poverty, with considerable variation across regions and over time. Second, data availability for India is excellent, allowing us to measure our explanatory and dependent variables with precision. Third, India's federal political structure allows us to explore variation across states within one national context. Finally, we have done much of our own fieldwork in this country.

Following the Indian case study, we conduct eleven smaller case studies across the world. We choose countries at different income levels, from different continents, and with different geographies and resource endowments. We also consider a priori information about variation in the explanatory variables of interest for our theory. Through this case selection process, we ensure that we have a diverse set of countries under investigation and can make broadly applicable claims about our theory. While our case selection is primarily informed by explanatory and control variables, we have found cases with considerable variation in the outcome. We also choose one case, Senegal, that initially presents a puzzle and a possible anomaly for our theory: the government's progress has been much faster in improving access to modern cooking fuels than in rural electrification.

We study each case in a structured manner, focusing on evaluating the values of the explanatory and dependent variables, along with some process tracing to check causal pathways and mechanisms. We emphasize that the primary goal of our qualitative analysis is not to correlate explanatory and dependent variables but to examine the ability of our theory to explain variation and outcomes within each case. Given the considerable differences in national circumstances across countries, we believe that the strongest evidence for or against our theory comes from a careful evaluation of the processes that have, or have not, led to the eradication of energy poverty over time.

This combination of methods inevitably leaves any one of them vulnerable to criticism. Our hope is that the methodological triangulation allows us to provide a totality of evidence that is compelling to scholars and practitioners with various backgrounds. The strength of the national case studies is their focus on the important decisions taken at the federal level, whereas the advantage of the subnational cases is their attention to detail and flexibility in the interpretation of processes, mechanisms, and junctures.

Implications

The large body of literature on energy poverty has so far had little to say about the politics of the problem (Thompson and Bazilian 2014). While the debate on effective policy has been intense (Barnes 2007; Reddy, Balachandra, and Nathan 2009; Nygaard 2010; Rehman et al. 2012; Coelho and Goldemberg 2013; Balachandra 2011b), this literature does not develop or test hypotheses about the political incentives that shape government policy. Stated bluntly, our study shows that much of the debate overlooks the root cause of bad policy.

On the one hand, governments with limited interest in reducing energy poverty would not act even if they knew exactly which policies would be effective, sustainable, and affordable. When studies investigate the policies of these governments and declare them ineffective based on poor outcomes (Gaunt 2005; Barnes 2011; Bazilian et al. 2012; Bhattacharyya and Ohiare 2012; Andadari, Mulder, and Rietveld 2014; Coelho and Goldemberg 2013), the research risks conflating the intrinsic de jure qualities of the policy with the government's indifference to competent implementation. Our findings show that the formalities of policy design are hardly relevant in the absence of strong government interest. Consequently, only cases that can demonstrate strong government interest are of use for policy analysis. Most studies to date ignore this basic point, and our analysis reveals the perils of this approach.

On the other hand, government interest is not itself a sufficient condition for success. Among governments with an interest in eradicating energy poverty, variations in institutional capacity and local accountability are important predictors of success and failure. When studies focusing on government interest (Min 2015) reveal a robust correlation between political incentives and the eradication of energy poverty, they point to a clear difference between countries with and without committed, interested governments. However, they do not explain why some countries with a high level of government interest make great progress in reducing energy access while others, despite an equally high interest, achieve little or nothing.

While the relationship between institutional capacity and policy quality is itself obvious, emphasizing institutional capacity is critical because a focus on policy quality masks the fundamental origin of the quality. Energy access policies can change rapidly, but they do not come out of the blue. The government's ability to formulate effective policies to end energy poverty depends on adequate institutional capacity, and such capacity cannot be built overnight. Government agencies evolve slowly over time, and often in a path-dependent manner.

Local accountability is a factor that most country-level case studies and cross-country analyses simply ignore. And yet, as Thompson and Bazilian (2014, 130) write, there are good reasons to consider local accountability essential for eradicating energy poverty: "A careful alignment of the various access points for determining energy policy with the broader more comprehensive contours of a nation's political geography would find that the strengthening of local institutions could provide the foundational seeds for broader national reform efforts—both in democratization and energy service provision." The identification of local accountability as a central factor in the eradication of energy poverty, both theoretically and empirically, thus adds a new dimension to the problem, breaking new ground and creating opportunities for future research.

In combination with the fundamental factors of energy resources, geography, and economic development, our analytical framework lays a solid foundation for energy access policy. Government interest, institutional capacity, and local accountability allow policymakers to identify the systematic causes of bad and good policy. Where the absence of these factors discourages effective energy access policy, the first priority is not to worry about the details of policy but to consider ways to increase government interest, build institutional capacity, and invest in local accountability. The identification of these systematic causes also opens a better opportunity to understand the effect of policy design on outcomes. That is, while energy resources, geography, and economic development are useful alternatives to explain energy poverty, they are insufficient for a complete explanation of the problem as there is considerable variation even after accounting for these factors. By accounting for the conditions that influence success and failure, scholars can identify the importance of policy design choices, all else constant.

Indeed, our analysis also offers a message of hope for the eradication of energy poverty. We show that after accounting for the fundamentals, governments' behavior and policies play an important role in determining progress in the eradication of energy poverty. Our analysis confirms that the increased interest in energy poverty among social scientists holds high promise for first identifying and then removing fundamental institutional barriers to progress. We show that otherwise similar countries have seen clearly diverging outcomes in mitigating energy poverty because of variation in the political context. Deterministic thinking about energy resources, geography, or economic development fails to explain much of the variation in energy poverty today.

Moving beyond the case of energy access, our results also have notable implications for the study of the provision of public goods and public service delivery in general. In comparative politics and comparative political economy, the literature on this topic is vast (Lake and Baum 2001; Adserà, Boix, and Payne 2003; Besley et al. 2004; Banerjee and Somanathan 2007; Besley, Pande, and Rao 2012; Tsai, 2007). While this research generates much knowledge and offers many valuable lessons about the topic, the studies typically fail to distinguish between different types of public goods and services. This generic, abstract approach is appealing from the perspective of theory building and external validity, but the downside of this appeal is the risk of obscuring important differences (Batley and Mcloughlin 2015). At worst, the lack of thorough comparative analysis of services may result in faulty inferences. As Kramon and Posner (2013) note in their review of studies of ethnic favoritism in public service provision, "Nearly all of them are vulnerable to a common and potentially devastating criticism: namely, that the pattern of favoritism that has been identified with respect to the outcome in question may be counterbalanced by a quite different, even opposite, pattern of favoritism with respect to other outcomes that are not being measured" (462).

We have shown the clear difference between government incentives to act on rural electrification and modern cooking fuels. While we have been able to apply the same theory to both issues, the outcomes are impossible to understand without careful attention to the technical details of the differing operational challenges, such as the network nature of the electric grid (Hughes 1993; Kale 2014a). The same distinctions probably apply to other public goods as well, ranging from road infrastructure to sanitation and the establishment of local health centers. By investigating these distinctions, we offer a solution to the question that Kramon and Posner (2013) raise for scholars of the politics of public services and public goods: If studies of distributive politics generate different results depending on the outcome one studies, how can scholars accumulate knowledge? We show that the solution lies not with aggregating results from various outcomes but in theorizing about how the intrinsic characteristics of different public services shape the politics of their provision. Aggregation of outcomes leaves scholars with a vague understanding of general tendencies, whereas comparisons of multiple goods and services allow genuine theoretical progress with precise testable implications.

In a broader sense, these findings suggest that there is merit to narrowing the range of social science theory and paying more attention to detail. While broad generalizations about factors such as regime type and all types of public goods can be elegant and profound, they can also hide important variation and, more important, stand between social science and actual efforts to improve policy. As the famed development economists Abhijit Banerjee and Esther Duflo (2014) put it in their advocacy of randomized controlled trials, "In the end, the choice facing the field of political economy is very simple. It can embrace grand theories that will offer us the satisfaction of strong and simple answers. Or it can try to be useful" (37). While we do not advocate for any particular method here or even criticize macrotheoretical analysis, we do agree with the importance of focusing on detail and not letting the pursuit of generality obscure important issues or stand between research and policy. This is important not only for development economists but also for political scientists, whose discipline is at the forefront of identifying and solving problems related to bad political incentives.

This discussion brings us to the importance of technology. With some exceptions (Cohen and Noll 1991; Skolnikoff 1993; Kim and Urpelainen 2013, 2014), scholars of political science have not given new technology the central role it should have in theory development and empirical research. Technological change is probably the most fundamental force that has shaped the evolution of human civilization over centuries (Basalla 1988), and the war on energy poverty is no exception. As we will see in many cases, such as Kenya and Bangladesh, new technology has both allowed governments to overcome their limitations and private entrepreneurs to substitute for uninterested or incapable governments. Had we not explicitly built technology into our analytical framework, we would have missed these important developments and failed to account for the surprising level of success in mitigation of energy poverty in these cases.

Finally, our combination of national and subnational case studies is itself noteworthy. We have reached beyond the typical approach of combining paired case studies at the national level. Specifically, one case study is a deep, longitudinal analysis of India's energy access from a political economy perspective; the other is a structured comparative analysis of eleven cases. We believe such triangulation is essential for studies that seek to tackle a new and technically complex topic.

Chapters of the Book

This book has seven chapters. Chapter 2 expands on the concept and reality of energy poverty. We provide precise analytical definitions of energy poverty and delineate the scope of our analysis, emphasizing that our focus here is on access to modern energy among the rural poor. We also review the history and contemporary situation of energy poverty at the global level, collating and summarizing the most recent data available. We provide a concise summary of the arguments found in the fragmented literature on energy poverty, emphasizing explanations for variation in the ability of countries to escape the energy poverty trap. Finally, we address the raging debates of the day, such as the relationship between energy poverty and climate change, providing concrete guidelines for dealing with the ethical dilemmas that efforts to make energy poverty history raise. In chapter 3, we present a general theory of energy poverty and variation in government efforts to combat it. The chapter explains how government interest, institutional capacity, and local accountability together allow countries to make rapid progress in alleviating energy poverty. We begin with a summary of our key assumptions and then proceed to derive empirical implications from them. Throughout the chapter, we illustrate the logic of the argument with examples, descriptive statistics, and narratives. The chapter also considers the role of new technology as an enabler of progress and contrasts the challenges of rural electrification and provision of modern cooking fuels.

Since much of our own fieldwork focuses on India, chapter 4 provides a detailed narrative of India's energy poverty trap and efforts to escape it. We show that India's national trajectory and differing rates of progress across states are both consistent with our theory. Since India's declaration of independence in August 1947, government interest has determined the broad pattern of rural electrification and progress in clean cooking fuels, but institutional capacity and local accountability have determined the rate of progress. The use of detailed household data on energy access shows that much of the variation in progress across Indian states can be attributed to these factors after we control for economic wealth, energy resources, geography, and other conventional explanations.

In chapters 5 and 6, we conduct eleven additional country case studies.[4] These case studies all follow the same research design and logic of inquiry, but they are not as detailed as the case of India. We chose countries from all continents, across a wide variety of wealth levels, and representative of various geographic contexts and resource endowments. We examine in detail the national processes that have resulted in success or failure of policy to eradicate energy poverty in terms of rural electrification and access to modern cooking fuels. These chapters complement the detailed but geographically restricted analysis of India's efforts to reduce energy poverty. Chapter 7 offers a concluding discussion in which we summarize our argument and explain its implications for the study of energy poverty and political economy more generally. We finish the chapter with a discussion of the implications of our findings for policy, emphasizing that our results highlight the centrality of policy for progress in rural electrification and provision of access to modern cooking fuels. The concluding chapter breaks new ground for research and practice on energy poverty as a policy problem, with an emphasis on previously neglected political economy considerations.

2 Understanding Energy Poverty in the World

Energy poverty is a multidimensional problem, if only because humans use energy for several purposes. The main uses of energy are lighting and appliances, cooking, heating, and transportation. Of these, we focus on two aspects of energy use that are directly related to everyday life and best reflect the idea of energy access as a basic necessity: access to electricity and clean cooking fuels in households. Transportation fuels are less of an issue for the very poor, as they do not own mechanized vehicles; heating is unnecessary for most of the year in many developing countries. Because the problem of energy poverty is more prevalent in the countryside of the nonindustrialized world, we train our analytical lens on rural areas in developing countries.

At the most fundamental level, household electricity is used for lighting and to power appliances. Lighting allows households to work, study, and play at night (Khandker, Barnes, and Samad 2009; Aklin et al. 2016). Traditional technologies, such as candles and kerosene wick lamps, provide an ineffective and unreliable lighting solution. Compared to them, electric lighting is more effective and reliable. Besides lighting, important electric devices include televisions, radios, mobile phones, fridges, freezers, and computers. In some countries, electricity is also used for heating and cooking. At the industrial level, electricity is a required input for most productive activities. Factories and commercial establishments need power to run machinery and appliances. In fact, our modern economies are unimaginable without reliable and affordable electricity.

And yet energy is more than just electricity. From an evolutionary perspective, cooking with heat has shaped the emergence of the human species itself. Some anthropologists have suggested (although this is the object of much debate) that the cooking of food was instrumental to the growth of the human brain because cooked food releases more energy than raw food does (Carmody and Wrangham 2009). Cooking can be done by different means, and the kind of fuel and the technology used for this activity has broad consequences. Today, experts commonly distinguish between traditional biomass—firewood, straw, agricultural residue, cow dung, and other organic material that can be burned for heat generation—and other fuels. A closely related common distinction is made between solid and nonsolid fuels. Some cooking fuels, such as coal and charcoal, are solid and require mining or processing for use. However, most modern cooking fuels are nonsolid (gas, electricity). For practical purposes, reliance on biomass using inefficient, traditional cooking methods is the central cause of energy poverty in cooking. Modern technologies can both improve the efficiency and reduce air pollution from cooking with biomass.

Historically, traditional biomass has been the most important source of cooking energy. For the longest time, firewood provided the necessary fuel to process meat and vegetables. The situation changed only in the early nineteenth century with the spread of gas stoves in the United Kingdom (Ravetz 1968). Today, "gas" encompasses natural gas, common in North America, and liquefied petroleum gas (LPG), a fuel that began to be used in countries such as Brazil in the 1930s (Jannuzzi and Goldemberg 2014; Poten and Partners 2003, 12). It is particularly popular in developing countries because it is easy to transport and use (Jannuzzi and Goldemberg 2014). It is also often more affordable than other modern fuels, especially electricity.

The continued use of traditional biomass for cooking is a problem because it is not only inconvenient; it also has serious negative health effects due to indoor air pollution (Barnes et al. 1994; Ellegard 1996; Bailis et al. 2009). When it is difficult to buy modern cooking fuels, people—usually women and children—have to collect wood and other solid fuels on their own (Goldemberg et al. 2004). This is time-consuming and reduces the number of hours available for school and productive activities. Access to convenient and clean modern fuels would allow more productive use of time and contribute to improvements in public health.

Energy is also needed for heating and cooling (Trace 2015, 160). From firewood to modern gas-based heating systems, a number of societies live in areas that require an elaborate and reliable heating system. Nonetheless, heating is not a major part of our story. The reason is that the poorest countries, in terms of both energy and income, tend to be located close to the tropics. Thus, the demand for heating tends to be fairly low among these populations. In contrast, air-conditioning and circulation, which are essential in hot and humid environments, require electricity. Besides being a source of discomfort, heat also reduces the life span of perishable food (Trace 2015, 164). Finally, transportation requires a lot of energy. Some transportation methods, such as trains, often rely on electricity or, historically, coal. Others, such as cars and planes, consume oil. Transportation currently represents about a quarter of energy demand in industrialized countries (Rodrigue and Comtois 2013). Just as electricity is essential for our economies, so is the ability to ship goods and people across long distances. Transportation costs are an important determinant of the ability of countries to engage in international trade, another source of wealth (Hummels 2007). Despite the importance of transportation, the cost and supply of this type of energy are often beyond the control of a single country. Oil prices are negotiated in global markets. Of course, policymakers can affect domestic transportation costs, but their ability to do so is limited, and the decision-making process operates in a manner that is different from our argument here.

To study energy poverty, we need a clear definition of what it is: *energy poverty means that a household is deprived of access to sufficient electricity or modern cooking fuels to meet basic household needs, or both*. This definition does not offer any particular threshold of sufficiency, but it emphasizes deprivation of access as the key issue. Because our goal here is to conceptualize energy poverty as a policy problem, it is not necessary to debate specific thresholds, though we will touch on existing proposals. One advantage of our definition is that it reduces difficulties in the measurement of energy poverty. As we have noted, energy poverty has multiple dimensions and goes beyond a simplistic dichotomy between the energy-poor and energy-rich. Several studies have sought to measure them individually (Pachauri and Spreng 2011; Aklin et al. 2016; Sadath and Acharya 2017; SE4ALL 2017a). Given the specificity of each country that we study in this book, our flexible approach relaxes the need to find strictly comparable measures.

Our starting point is thus that people have basic energy needs to be filled. By "sufficiency," we also mean access at times of need. For example, intermittent electricity supply or unreliable access to LPG for cooking would compromise sufficiency (Aklin et al. 2017). If people do not have access or the means to purchase the energy they need, they are defined as energy-poor. This view is very similar to the one adopted by international organizations such as the International Energy Agency (IEA 2010b, 2014a).[1] As soon as these two conditions are met, a household is not considered energy-poor, regardless of the context. Energy needs may vary depending on factors such as temperature and culture, but the combination of access and affordability is the lowest common denominator for escaping energy poverty in any society.

In the next section, we explain what energy poverty means in practice and which conditions need to be met for a household to be lifted out of energy poverty. Before turning to these issues, however, we note that energy poverty has two forms. First, a household is energy-poor if it does not have physical access to energy. One reason could be that no supplier is willing to deliver LPG in a particular region or the electric grid does not extend to a given village. Second, a household is energy-poor if the price of energy is too high for the consumer. There is, of course, a degree of arbitrariness when one has to judge whether energy is too expensive, but when millions of people lack such a basic necessity, a mismatch between supply and demand becomes clear. In a general sense, both circumstances can be combined into the idea that energy poverty occurs when modern energy supply does not match demand because the latter is not important enough, be it financially or politically.

An alternative approach from the literature defines energy-poor households based on the share of their income devoted to energy. This view relies on the stylized fact that poorer households spend a larger share of their income on energy (Leach 1987). The British government, for instance, has defined an energy-poor household "as one which needs to spend more than ten per cent of its income on all fuel use and to heat its home to an adequate standard of warmth" (cited in Pachauri et al. 2004). However, the use of a fixed threshold is problematic for two reasons. First, this would necessitate using a largely arbitrary number. Why is 10 percent better than, say, 15 percent? Motivating a single number for the entire world is too ambitious a task. Second, there are systematic variations across countries in energy costs (Foster 2000). The threshold may be country dependent for both exogenous or endogenous reasons. Therefore, although we consider affordability in our definition, we do not propose a fixed threshold.

Another existing approach is to define energy poverty in terms of the technology that people use. For example, a household using energy sources such as animal dung or wood would qualify as energy-poor (Sagar 2005). This approach is useful for cooking fuels because the primary problem lies with reliance on traditional biomass, but it is not entirely appropriate for electricity access. While electricity access itself is clearly a superior solution to traditional alternatives such as kerosene lanterns, the fuel used to generate electricity is itself not important. Given these reasons, we rely only partly on technology as an indicator for modern energy access.

Khandker, Barnes, and Samad (2012) offer a flexible approach to determining the energy poverty status of a household. Drawing on Indian data, they note that up to a certain level of income, typical energy consumption (measured in physical units) does not vary much. They thus define a household as energy-poor if it cannot consume this typical energy consumption level. The major advantages of this flexible approach are that it can be applied in different social contexts and can be measured with standard household surveys. The approach is in spirit consistent with ours, but in our empirical analysis, we cannot apply it because we lack the detailed survey data necessary for most countries in the world.

To set the stage for our analysis, the next section discusses the nature of energy poverty and the reasons for its persistence. Then we present a global overview of energy poverty today, describing differences within and across regions. The remaining sections provide an account of existing explanations for energy poverty, highlighting their strengths and weaknesses. In the final section, we describe the global policy debate on the relationship between sustainable development and energy poverty.

Energy Poverty: Problems and Solutions

The problem of energy poverty is much more serious in rural than in urban areas, as providing energy to urban areas is easier than addressing rural energy poverty (Barnes 2014). In cities and towns, economies of scale from high population densities facilitate providing energy to a large number of people. For instance, once a power station is built and an urban center is connected to it, providing electricity to the rest of the area is straightforward.[2] In contrast, rural population densities are so low that electrifying villages, not to mention households, is a major challenge. As the distance between existing electricity infrastructure and villages grows, the cost of the work increases. Furthermore, electricity losses from transmission and distribution mean that the amount of electricity consumed is less than the amount electricity produced. For example, Barnes (2014, chap. 2) notes that the last-mile distribution of electricity to remote rural areas is a particularly difficult challenge.

Policymakers must weigh whether these costs are exceeded by the benefits of energy access. Therefore, we need to examine in a concrete sense the challenge of energy provision in rural areas. Let us first consider the case of electricity. In developing countries, household electricity is used for lighting and electric appliances among the rural poor. Lighting is useful for both leisure, such as reading, and productive work. For instance, outdoor cooking is much more convenient with electric lighting than with candles. Electric lighting enables people to be free to decide when they want to cook instead of cooking only during the day, when there is sunlight. Lighting is also invaluable for small businesses that operate at night.

In the absence of electric lighting, many households rely on inferior alternatives, such as candles or kerosene lamps. These solutions are often impractical because the quality of lighting from these sources is usually poor. Sometimes it is also costly, in terms of both time and money, to get adequate supplies. Kerosene fires also are a constant risk in households. Finally, these solutions are a threat to human health. Kerosene lamps are a source of lung cancer and respiratory illnesses; unsurprisingly, they are mostly used in the poorest countries (Apple et al. 2010; Lam et al. 2012). Electrification is thus not only a motor of economic growth but also has notable public health benefits.

According to our definition of energy poverty, there is a problem if a household lacks access to a reliable source of electricity. One reason could be that the household lives in a nonelectrified village: if the grid is not extended to the community, there is little the household can do to be connected. Affordability can be a problem that originates from both producers and consumers. Given that the deployment of both grid and off-grid electricity is capital intensive, the lack of electricity access can be caused by prices being set too high or consumers being too poor. In practice, a government may be able to adapt costs based on a group's income level. For instance, the very poorest villages could receive subsidies while the price would be set at the marginal cost of production in wealthier regions. In this sense, the problem with lacking electricity access mostly originates on the production side: government's commitment to helping the poor through electricity subsidies or providing them with cash transfers that will allow them to buy electricity.

In practice, access to electricity can be provided through the grid or from off-grid systems. Grid electricity is powered by coal plants, hydropower from rivers or dams, nuclear plants, and so forth. The idea of off-grid power is often associated with renewable energy, such as solar microgrids, but it can also be powered by fossil fuels such as diesel. Historically, industrialized countries have typically relied on the grid to provide electricity (Smil 2010). Some developing countries, such as China and Vietnam, are following a similar path; others, such as Bangladesh, have invested heavily in off-grid solutions.

An electricity network has three components. First, a source of power must generate electricity. The most common way to do this is to use a thermal plant. A fuel such as coal is used to heat water, and the steam then spins a turbine connected to an electric generator. Various fuels can perform this task: coal, oil, or uranium. Alternatively, hydropower follows a similar process, but the fuel here is flowing water. The flow can be induced by a dam or can originate in a natural setting, such as the stream of a river. Historically, the ability to generate electricity depended on a country's natural endowment. The United Kingdom, for instance, had ample reserves of coal that facilitated the Industrial Revolution.[3] As markets grew more connected over the nineteenth and twentieth centuries, the lack of domestic resources became less of a problem because coal, oil, and other fuels could be bought from global markets.

A country's electricity mix is important because sources of power work in different ways. A nuclear reactor produces a stable flow of electricity that is available at almost all times. The disadvantage is that turning off a nuclear power station is cumbersome, meaning that a plant manager would rather pay people to use its electricity than turn it off. Other sources, such as hydropower, are convenient because they are both reliable and can be used on demand except when deficient rainfall creates problems.[4] Finally, new renewable energy sources such as wind and solar power tend to depend on external factors such as the strength of winds and solar radiation, although the development of batteries somewhat reduces the problem. A reliable electricity system must thus have enough power to cover the base load requirement (minimum demand) and be able to respond to unexpected increases in demand (peak demand). Countries with reliable electricity systems tend to do so by using a variety of complementary sources.

The second component is the transmission and distribution of electricity. Power must be transmitted from the power source to the consumer in two steps. First, electricity goes to electrical substations, which reduce the voltage and distribute electric power to households and other consumers. This is a critical stage in many developing countries, where not all the electricity generated by power stations ultimately reaches the consumer. A notable fraction of it is lost due to distance and the quality of the transmission design. For instance, electricity losses increase with distance between the power station and the consumer. Similarly, losses tend to be lower for underground transmission systems, though they cost more to construct. In the United States, this loss is estimated to be about 5 percent, but developing countries suffer from much higher losses: in low- and middle-income countries, the average losses were above 10 percent in 2014.[5] Another fraction of it is lost due to theft, a major issue in many developing countries. Instead of paying a connection fee, people illegally connect their own household to the grid. Estimates of theft vary, but experts believe that it can climb to about 14 percent of total generation (Smith 2004, 2070). Theft increases with corruption, further emphasizing the role of good governance for the provision of electricity (Smith 2004).

Transmission is critical in two additional ways. Besides economies of scale for transmission in high-density areas, there can also be economies of scale if a productive load is already allocated to a factory or some other big consumer. For instance, if electricity is transmitted to, say, a milk factory in a rural area, adding lines to neighboring private households does not incur a high marginal cost. Furthermore, transmission is critical when suppliers need to decide which customers to prioritize. If power is scarce, utilities or the officials in charge may have preferences for specific regions. Thus, power may be systematically diverted to some consumers.

The last component of an electric system is the connection of the household to the grid. Some countries have reasonably good and well-managed networks but come up short when trying to actually connect people to the grid. An example is Ethiopia. According to a World Bank official, "Ethiopia has a competent public utility, with a mandate to increase the number of rural towns and villages with electricity access; and government and donor support for the rural electrification program, including through IDA [International Development Association]. But a few years ago, when we visited some towns and villages that had been electrified for 18 months, we were shocked to see that the program had not gone the 'last mile' to ensure that households were actually connected to the grid."[6] One cause of this problem is that connecting a household is the first time a supplier can get money from consumers, and thus connection costs are often high—indeed, too high for many poor rural households (Golumbeanu and Barnes 2013). According to the same report, the connection cost in Ethiopia rises up to 15 percent of a poor household's annual income, a huge cost that undoubtedly encourages theft.

For off-grid systems, the problem is both similar and different. It is similar in the sense that an off-grid system, such as a microgrid, requires a power source and distribution lines to each household. It is different because the scale of such a system is typically fairly small. Some firms sell their systems, and others rent them. Regardless of the business model, many of these firms are run privately, though in many countries, they receive some form of support from the government (Palit and Chaurey 2011). At any rate, the capital investments and technological know-how required to deploy off-grid systems are lower than in the case of a national grid. In theory, off-grid systems are therefore flexible and can be installed and maintained on demand. In practice, things are more complicated. Customer service is often lacking, meaning that a technical issue can linger for a long time. In fact, one of the strengths of off-grid systems, insulation from external changes in consumption, is also a weakness: providers do not always feel the pressure to respond quickly to technical issues. Consumer satisfaction suffers, leading to distrust toward the off-grid technology (Aklin et al. 2014a).

Overall, then, constructing a functioning electric system requires sound public policy. Available evidence suggests that the quality of electricity supply is highly correlated with the quality of governance (Smith 2004). Some governments have been successful, but many have failed. Estimates vary, but it is commonly admitted that more than 1 billion people do not have access to electricity in their home, and many more have only an unreliable supply of power (Birol 2007; IEA 2017).

What about cooking? Energy poverty with respect to cooking is defined as the inability to cook with convenient, modern, nonsolid fuels. Energy-poor households thus rely, for example, on wood, crop residues, dung, and charcoal (Bonjour et al. 2013). These fuels are often used to provide heat through an outdoor open fire or a simple stove, possibly located indoors. In contrast, nonsolid fuels include electricity, natural gas, LPG, and biogas. Cooking based on solid fuels is technologically far less demanding than cooking based on nonsolid fuels. It is also often cheaper, because solid fuels are often available for free and basic stoves are easy to make. However, solid fuels are both inconvenient and present a host of environmental and public health problems (Ellegard 1996; Schlag and Zuzarte 2008; Wickramasinghe 2011; Bansal, Saini, and Khatod 2013; Goldemberg et al. 2004). Modern cooking fuels, such as LPG, are safer to use as they reduce the risk of home fires (Kandpal, Maheshwari, and Kandpal 1995; Lucon, Coelho, and Goldemberg 2004; Quaye-Foli 2002). In its 2017 *World Energy Outlook*, the IEA estimates about 2.8 million premature deaths around the world from indoor air pollution due to the use of solid biomass for cooking (IEA 2017, 14). Based on their "energy access for all" scenario, an estimated 700,000 people will still die prematurely by 2030 (IEA 2017, 72).[7]

Solving cooking fuel poverty consists of two steps. First, a household must have access to modern cooking fuels. In many industrialized countries, cooking is commonly done by using electricity or natural gas. However, LPG is more often found in developing countries. None of these fuels require households to expend time and effort on collection. They also generate virtually no indoor pollution and are thus superior from a health perspective. Furthermore, they are also more efficient technologically. The ratio of energy conversion from solid fuel is quite low; an open fire uses only about 5 percent of the energy that is liberated by the fire (Smil 2010). That number is significantly higher for nonsolid fuels.

Unfortunately, access to modern cooking fuels is not easy to guarantee. While solid fuels can be collected locally, the deployment of nonsolid fuels requires coordinated and well-organized suppliers. If a country wishes to use electricity to power its stoves, it must overcome the obstacles to constructing an efficient electricity network.[8] Similarly, natural gas requires extended networks of well-maintained pipelines. A difficult task even in a single city, it can become extremely costly if remote villages have to be connected. At the national level, a country that wants to use gas for cooking needs to draw on its own reserves or have access to global gas markets.

Of course, less complicated solutions do exist. One reason that LPG became very popular in developing countries is that it does not require the same complexity of infrastructure. LPG is distributed through cylinders that can be transported in trucks, and no pipelines or power transmission lines are needed, though facilities such as fuel depots are still necessary. LPG thrives thanks to the light infrastructure required. However, this does not mean that LPG is cheap and widely available (Schlag and Zuzarte 2008). Remote villages are still at a disadvantage; they can often barely afford LPG, meaning that in the absence of subsidies, few suppliers have an interest in supplying them. Furthermore, even though the kind of infrastructure for LPG is less complex than the one needed for electricity or natural gas, it is still an infrastructure. Trucks must be able to reach remote places, which can be difficult in developing countries. Roads are often in poor condition and regularly become useless because of flooding during the rainy season (Mwabu and Thorbecke 2004). Therefore, solving cooking fuel poverty requires the development of reliable infrastructure. This infrastructure can be as elaborate as the one required to improve access to electricity. However, it can also be lighter, an option that is not available for electricity.

The second step is to equip households with stoves that are compatible with modern fuels. The cheapest suitable stoves can cost less than US$100, whereas larger and more complex ones can cost up to US$500.[9] Stoves differ in how convenient they are for different kinds of food, their weight, whether they can be assembled on the spot, and how much food they can cook at the same time. To some degree, their deployment depends on their price as well as how adapted they are to local demand.

Modern cooking stoves are often too expensive for the poor. For instance, Schlag and Zuzarte (2008, 12) contend that the cost of adopting stoves has been a major obstacle to their deployment in sub-Saharan Africa. It is unsurprising that households are reluctant to switch to modern cooking technologies if the fixed costs are high and the marginal costs not particularly competitive either. Furthermore, modern cooking stoves were first developed in industrialized countries with little regard for the preferences of the end users (Barnes et al. 1994, 14). Despite the theoretical benefits of these stoves, their adoption is likely to remain limited if they do not respond to demand and adapt to local needs. Finally, modern stoves require people who know how to maintain and repair them. Users can be trained, but more often than not, some form of customer service is needed to ensure that consumers can receive help if there is failure. In the absence of a reliable support system, people may be reluctant to spend their savings on these new technologies.

Overall, cooking fuel poverty can be analytically decomposed in two problems: capital investments in modern stoves and building infrastructures for the distribution of the corresponding fuel. If the fixed cost of buying a modern stove or the marginal cost of fuel and maintenance are too high for a household, then we consider it to suffer from energy poverty with respect to cooking. Of course, there are other reasons households might not use modern cooking fuels, such as cultural ones. We readily admit that these are real issues that have considerably slowed the deployment of clean and modern cooking technologies, but we also note that our definition of energy poverty focuses on the financial and physical obstacles to cooking fuel and devices. We believe that the vast majority of the world's population would welcome affordable, convenient, and safe cooking fuels. With natural gas, petroleum, or LPG gas being among the cleaner and more efficient cooking fuels on the "ladder of fuel preferences" (Leach 1987), these versatile alternatives are often not adopted for mainly two reasons: fuel cost (Schlag and Zuzarte 2008; Viswanathan and Kumar 2005) and the incompatibility of modern fuels with existing biomass stoves (Levine and Cotterman 2012; Miller and Mobarak 2011; Mobarak et al. 2012).

A Historical and Contemporary Overview

Low electrification rates, lack of or unreliable access to the national grid, high connection charges, and limited access to modern cooking fuels such as LPG are only some examples of the problem of energy poverty. Given that effectively addressing energy poverty is not a new challenge, we review how energy transitions were managed in today's industrialized countries. Only then do we map the current landscape of energy poverty on a global scale. We show that there is not only considerable variation in energy poverty across different countries and regions, but also within most regions, not to mention most countries.

Many developing nations today face the challenge of providing citizens with reliable and permanent access to electricity and modern cooking fuels, but industrialized countries managed to solve this basic problem almost a century ago. This transformation, often referred to as the third energy transition (Smil 2010), was a difficult and slow process (Fouquet 2008, 2010). Spearheaded by the United Kingdom's transition from traditional energy sources to fossil fuels at the onset of the Industrial Revolution in the late eighteenth century, Fouquet (2010, 6586) identifies "the opportunities to produce cheaper or better energy services" as the key driver underlying any energy transition. While new energy services initially come with a price premium that limits their early use to niche markets, prices fall over time as economies of scale grow and efficiency increases. Ultimately, technological progress crowds out obsolete energy services. In the beginning, the use of steam engines, for instance, was limited to coal mines because these engines were a good way to transport water out of the mines, which made coal production cheaper. Because of their use in coal production, the fuel efficiency of steam engines increased, starting their triumph in the textile industry, in which it replaced humans and animals for power production (Fouquet 2008).

These transitions often span several decades or centuries, as in the case of coal replacing biomass as the primary energy source in the United States (Schurr and Netschert 1960), transitions to electricity (Fouquet 2010), and the increased use of LNG (Smil 2010). Although the basic dynamics of these transitions are similar across different sectors such as power, heating, transport, and light, there is considerable variation in the technologies used and the institutional contexts. Energy transitions are complex trajectories that depend on historical developments, resource availabilities, demand for modern energy services, and supporting policies for phasing in new technologies and practices (Gales et al. 2007; Madureira 2008; Fouquet 2008). These historical lessons show that both the complexity and slow speed of energy transitions are factors in understanding energy poverty in the developing world.

With these historical energy transitions in mind, it is not entirely surprising that energy poverty continues to be a serious problem for many countries. The IEA (2017, 11) estimates that almost 1.1 billion people in the world lack access to electricity. Virtually all of these people today live in sub-Saharan Africa or the developing countries of Asia. According to recent data, about 590 million people in sub-Saharan Africa (57 percent of the regional population) and 440 million people in developing Asia (11 percent of the regional population), respectively, are lacking electricity access. The use of traditional biomass is an even more common problem;

the IEA (2017) estimates that as many as 2.8 billion people continue to lack access to modern cooking fuels. Here, the massive population of Asian developing countries leaves 1.65 billion (43 percent of the regional population) reliant on traditional biomass; in sub-Saharan Africa, 780 million (80 percent of the regional population) are in the same situation. While there has been progress over the past ten to fifteen years, at least for electricity access, the newest *World Energy Outlook* scenarios still suggest that energy poverty will remain on political and development agendas for years to come. By 2030, the IEA (2017, 48) expects that 674 million people will still remain without electricity access, most of them in rural sub-Saharan Africa. For access to modern cooking fuels, the outlook remains even more skeptical, with an estimated 2.3 million people (down from 2.8 million in 2015) being dependent on biomass and other solid fuels for cooking (IEA 2017, 67).

More relevant than these aggregate figures is an accurate understanding of variation across different countries and regions. As the numbers already show, sub-Saharan Africa performs poorly across the board, whereas the developing countries of Asia have made much more progress in electrification than in the provision of modern cooking fuels. To map the energy landscape and illustrate global variation in energy poverty in a more precise manner, we need reliable data. Unfortunately, such data are difficult to obtain on a global scale. Notwithstanding the limitations, we can learn a lot from information from the Energy Access Database that accompanies the IEA's annual *World Energy Outlook*, the most comprehensive data source on energy access. It brings together multiple data sources, such as IEA energy statistics, the World Bank's Living Standards Measurement Surveys, and the World Health Organization's Global Health Observatory.[10] Leveraging and consolidating all of these different data sources, the database is a particularly powerful resource. It contains information not only on electricity access but also on access to modern cooking fuels. Therefore, the database reflects the two key dimensions of our notion of energy poverty.

The *World Energy Outlook* methodology defines *energy access* as "a household having reliable and affordable access to both clean cooking facilities and to electricity, which is enough to supply a basic bundle of energy services initially, and then an increasing level of electricity over time to reach the regional average" (IEA 2017, 21). While this definition excludes, for instance, electricity access to private enterprises or important buildings for public life, such as schools or hospitals, it emphasizes that energy access requires some actual threshold consumption of electricity and the provision of "cooking facilities which can be used without harm to the health of those in the household" (IEA 2017, 59). In this regard, this definition of energy access is similar to our broad understanding of energy poverty, which emphasizes access, reliability, and affordability.

This conceptual congruence notwithstanding, the actual measurements provided by the *Outlook* are based on simple binary measurements of whether electricity access is available to a household and whether traditional biomass is used as the primary cooking fuel (IEA 2014b, 3). Although the IEA strives for a nuanced conceptual understanding of energy poverty, its cross-national comparisons are, because of data limitations, based on simple measures of access to electricity and modern cooking fuels. Keeping these data limitations in mind is important, because the data we report are likely to underestimate the significance of the energy poverty problem. Households that are officially connected to the grid may not have reliable access to electricity. Power could, for example, be available only intermittently at particular hours of a day. It could also be too expensive for households to consume. Although we emphasize reliability and affordability of energy access in our working definition of energy poverty, we ignore them here because of the lack of detailed data.

Despite its limitations, the IEA database can reveal important insights when we treat electrification rates as a conservative lower estimate of the severity of the problem. Moreover, it is useful as a means to compare electrification rates across countries and regions. For this purpose, figure 2.1 shows electrification rates measured as the percent of households connected to the national electricity grid. These measures include both urban and rural electrification rates, but the problem of low electricity access is virtually always driven by the lack of household connections in the countryside. Except for Libya, for which we use 2010 data (99 percent electrification rate), we consistently use the 2016 data from the most recent *Energy Access Outlook* (IEA 2017, annex A, table A.1). Because energy poverty is most prevalent outside the Organization for Economic Cooperation and Development (OECD) member states and former Soviet states, which all achieved universal electrification decades ago, we do not include them in the data description.[11]

Electricity Access around the World

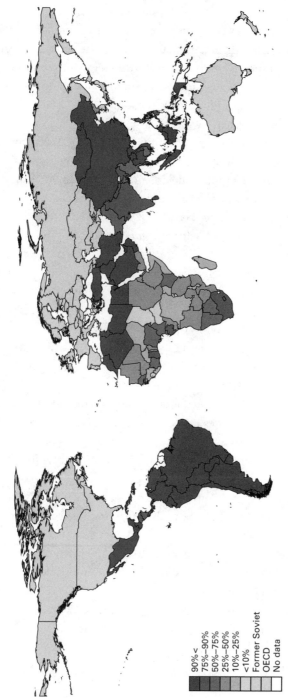

Figure 2.1

Electrification rates around the world. We do not report data for industrialized or former Soviet countries. For other countries, darker colors translate into higher electrification rates, the desirable outcome in terms of progress in addressing energy poverty. Data from 2016 or 2015, depending on latest observation available. Sources: IEA (2017); SE4ALL (2017a).

The overall pattern of electricity access shows considerable variation in electrification rates across the globe, even among developing and least developed countries. While electrification rates in Latin America and the Middle East are consistently well above 90 percent, reflecting high economic development and an abundance of natural resources, such as oil in the Middle East and hydropower in South America, the situation in South and Southeast Asia is much worse. There is also a remarkable divide between resource-endowed and relatively advanced economies in North African states, such as Egypt and Tunisia, and the least developed countries in sub-Saharan Africa, for which electrification rates hover at around 40 percent. Energy poverty, measured here as electricity access, thus approaches abysmally low levels in the least developed countries in sub-Saharan Africa.

Next, consider variation at the regional level. Grouping our countries of interest into five world regions according to the official World Bank classification (East Asia, Latin America, Middle East and North Africa, South and Southeast Asia, and sub-Saharan Africa), figure 2.2 shows box plots of electricity access by region. We report for each region the median electrification rate and the interquartile range, with the whiskers extending out 1.5 times the size of the box plot. Any data that fall outside this range are also added to the plot as individual points. The right-most column shows the box plot for the full sample across all five regions, again excluding OECD countries and former Soviet states.

As the figure shows, there is variation within regions: most countries in East Asia, such as China (100 percent) and Malaysia (99 percent), have achieved very high electrification rates, whereas electricity access in North Korea is limited to only one in four households. These differences are not too surprising given the large discrepancies in income levels, economic production, and the integration of the Chinese and Malaysian economies into the world market. Economic development is highly correlated with high levels of electrification, which is also evident in the context of Latin America. Except for Haiti (33 percent), a historically poor country and only slowly recovering from a devastating earthquake in 2010, all Latin American countries (average electrification rate of 93 percent) have typically high levels of electricity access. This fortunate situation results in part from economic development, but the widespread use of hydroelectric power is also relevant here. The abundance of oil and gas reserves in the Middle East

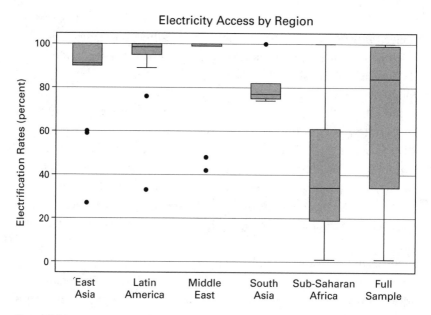

Figure 2.2
Box plot of electrification rates by region. The boxes show the interquartile range from the 25th to the 75th percentile, with horizontal lines indicating the median value. The whiskers extend up to 1.5 times the size of the box if such observations exist. Observations (i.e., countries) that fall outside this range are shown as dots. Data from 2016 or 2015, depending on the latest observations available. Sources: IEA (2017); SE4ALL (2017a).

underlies the almost universal electricity access in this region, with poor Yemen (48 percent) and Djibouti (42 percent) being the two exceptions. Saudi Arabia, for example, as a country with one of the largest oil reserves worldwide, has long been able to provide electricity to its 30 million inhabitants due to the gigantic revenues from oil exports that have allowed the government to invest in generation, transmission, and distribution.

The distribution of electricity access in South and Southeast Asia is fairly even around the mean access rate of 84 percent, with Nepal (77 percent), India (82 percent), and Pakistan (74 percent) all playing in the same league. Thailand and Vietnam have achieved electrification rates above 95 percent, while Myanmar remains at 59 percent and Bangladesh at 75 percent. These differences are particularly striking because they cannot be explained with reference to economic development only. In particular, Vietnam's impressive and early achievements in rural electrification cannot be explained with reference to wealth because rapid progress began decades ago when the country was still poor and had achieved little in terms of industrialization.

Variation in sub-Saharan African states is particularly extreme and virtually spans the entire range, with the vast majority of states grouping at the lower end. The only remarkable exceptions to this trend are Mauritius (100 percent) and South Africa (86 percent). While South Africa is the economic powerhouse in sub-Saharan Africa, Mauritius excels on several dimensions in terms of both economic development and good governance. In addition, the Mauritian government has a long-term energy strategy in place that aims to meet 35 percent of the country's energy demand in 2025 by renewable sources. Many of the other African countries in the sub-Saharan region fall well below the overall sample median value of about 34 percent of electrification access, which is largely a result of low development in this region. Access rates are exceptionally low, for example, in South Sudan (1 percent), the Central African Republic (3 percent), Sierra Leone (9 percent), and Chad (9 percent). All in all, the data clearly show variation not only across regions but also within regions themselves.

Yet another form of variation to consider is that between urban and rural areas. Since provision of electricity benefits from centralized production due to economies of scale, it is much easier and cheaper to achieve in densely populated urban areas, while rural electrification poses a particular challenge to eradicating energy poverty. Fortunately, the Energy Access Database also provides information on urban and rural rates of electrification, even though the IEA (2014b) technical report on the methodology used to construct the data points out that these comparative urban-rural electrification data largely result from secondary analyses and econometric estimation techniques based on socioeconomic, population, and energy consumption data.

These caveats notwithstanding, urban electrification rates are much higher than their rural counterparts. Even among sub-Saharan countries, the situation in urban centers is remarkably different from the countryside. Economic hot spots and centers for business and commerce, such as Nairobi in Kenya and Dar es Salaam in Tanzania, main transit points for all of East Africa, have electrification rates of 76 percent and 65 percent, respectively, while country averages are much lower, with about 60 percent in Kenya and less than one-third in Tanzania. This pattern also holds in many other African states, such as Cameroon, Ethiopia, and Zimbabwe, where the national capitals are relatively well electrified. The difference stems from several sources: electrification in cities is not only much easier, but capital cities are also the homes to the political leaders and the country's political, business, and military elite. Perhaps the presence of powerful people provides governments strong incentives to offer considerably better access to electricity, at least relative to the rest of the country.

Given comparatively low national electrification rates in many countries, the flip side of relatively high urban electricity access is that access for rural populations is even worse than the aggregate statistics suggest. While there is urban bias across the board, it is especially pronounced in sub-Saharan Africa, where the difference between urban (68 percent) and rural (14 percent) median electrification rates is more than 50 percentage points. The relative ordering of world regions in terms of how well they fare in addressing rural electrification is the same as for overall electrification. While sub-Saharan countries are doing the worst, South and Southeast Asian states occupy the middle ground, and the Middle East and Latin America achieve high rates of rural electrification, with median values of 99 percent and 91 percent, respectively. This is largely for the same reasons as already discussed, but one may add that rapidly developing countries such as Brazil, for instance, highly prioritize rural electrification in order to demonstrate global leadership and modernization.

Energy poverty is not only about electricity; it also reflects the paucity of modern cooking fuels. While data on the composition of primary cooking fuels on a global scale do not exist, country-specific evidence is available. Viswanathan and Kumar (2005, 1023) show that "firewood and dung present the main cooking fuels in rural India," while "the penetration of LPG is very weak with only 4 per cent of expenditure going toward this 'clean' fuel." The situation is admittedly precarious in rural India, with urban households spending a tenfold of the rural figures on LPG, but the predominant use of traditional biomass reflects the overall picture pretty well. Schlag and Zuzarte (2008, 3) find that more than 70 percent of total households in Tanzania, Kenya, and Malawi are dependent on biomass for cooking, with numbers rising above 90 percent for rural sub-Saharan countries. In many developing countries, charcoal is now used for cooking, which is more efficient in terms of energy content and also easier to transport. LPG, a mixture of propane and butane, is often considered the most efficient cooking fuel. Except for Senegal's butanization program, LPG use is impeded by high prices in countries like Kenya or Tanzania (Schlag and Zuzarte 2008). Reflecting common difficulties in transitioning to more efficient fuel use, the Senegalese government paired subsidizing LPG provision with the introduction of a new LPG-compatible cooking stove; this dual strategy proved immensely effective, increasing LPG use to 70 percent across the country and up to more than 90 percent in Dakar, Senegal's capital (Fall et al. 2008).

These individual cases are insightful, but they do not enable a systematic assessment of access to modern cooking fuels around the globe. To provide a comprehensive overview, we rely on 2015 data from the same *Energy Access Outlook* (IEA 2017, annex A, table A.2).[12] It provides information about what share of a country's population is without access to modern cooking fuels, which is defined in the following way: "*Access to clean cooking facilities* means access to (and primary use of) modern fuels and technologies, including natural gas, liquefied petroleum gas (LPG), electricity and biogas, or improved biomass cookstoves (ICS), as opposed to the basic biomass cookstoves and three-stone fires used in developing countries" (IEA 2017, 21). Importantly, these data are superior to data on traditional biomass use for cooking, which would gloss over potentially meaningful country differences across modern cooking fuels.

Figure 2.3 shows access to clean cooking fuels around the world, where darker-colored countries grant better access to modern cooking fuels, whereas lighter-colored ones rely heavily on traditional biomass as primary cooking fuel, such as firewood, charcoal, or crop residues. The main message from the figure is that the two components of energy poverty that we identify, access to electricity and access to modern cooking fuels, are highly correlated: countries with high electrification rates also have almost universal access to modern cooking fuels—for instance, Brazil (95 percent), which rolled out an extensive LPG program starting in the early 1950s (Lucon, Coelho, and Goldemberg 2004), and Saudi Arabia, which almost exclusively uses LPG and also has a large gas-gathering and processing system in place, run by ARAMCO, Saudi Arabia's national petroleum and national gas company. As Schlag and Zuzarte (2008) suggest, traditional biomass is used a lot in sub-Saharan African countries, such as Ethiopia (6 percent), Rwanda (5 percent), and Uganda (5 percent), where access to modern cooking fuels is minimal. In fact, in twenty of the forty-seven countries in our

Access to Clean Cooking Fuels around the World

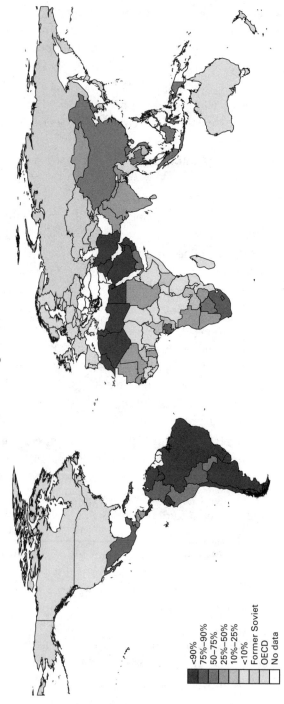

-90%
75%–90%
50%–75%
25%–50%
10%–25%
<10%
Former Soviet
OECD
No data

Figure 2.3

Access to clean cooking fuels around the world. We do not report data for industrialized or former Soviet countries. For other countries, darker colors translate into greater access to modern cooking fuels, the desirable outcome in terms of progress in addressing energy poverty. Data from 2015 or 2014, depending on the latest observations available. Sources: IEA (2017); SE4ALL (2017a).

sample of the sub-Saharan African continent, less than 10 percent of households have access to modern cooking facilities.

While the main patterns of variation in access to modern cooking fuels largely mirror what we have seen for electricity access before, variation within regions is even more pronounced than in the case of electrification rates. This is not too surprising because advancing access to modern cooking fuels often does not rank as highly on governmental agendas as does pushing for electrification. Disaggregating our sample into the same five world regions as above, the box plot in figure 2.4 verifies lacking access to modern cooking fuels in the poor countries of South and Southeast Asia (36 percent) and sub-Saharan Africa (11.5 percent). Substituting away from traditional biomass is appealing only if households can afford the change (Maliti and Mnenwa 2011; Rehman et al. 2005; Coelho and Goldemberg 2013). Emphasizing this point, the majority of East Asian and Latin American countries have good access to modern cooking fuels, while outliers in these regions, such as Myanmar (6 percent), Laos (5 percent), and Haiti (7 percent), are also among the poorest and less advanced economies. Because of their oil wealth, Middle Eastern states have long since replaced the use of inefficient biomass with liquid fuels and LPG in particular.

To summarize, the two dimensions of energy poverty that we emphasize are highly but not perfectly correlated. Countries that have achieved high levels of electrification are typically also better at providing their populations with access to modern cooking fuels. Because electrification is a much more salient issue in many countries' political agendas, however, the relationship between electrification and modern cooking fuels is not ironclad. Achievement in electrification is more impressive than achievement in the provision of modern cooking fuels, and we will show that policy plays a key role in producing this outcome. According to the IEA's latest *Energy Access Outlook*, about 600 million people had electricity access in 2016 who did not have connections in 2000, while access to cooking fuels remained virtually unchanged over the same time period (IEA 2017, 11–12).

Furthermore, there is considerable variation in energy poverty along three core dimensions: between regions, within regions, and between urban and rural areas within nations. High levels of energy poverty persist, and the situation is especially severe in sub-Saharan Africa and the least developed countries of South and Southeast Asia. Countries in the Middle East, Latin America, and North Africa fare much better. In general, rural

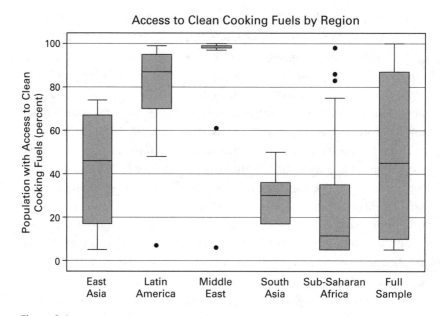

Figure 2.4
Box plot of access to clean cooking fuels by region. The boxes show the interquartile range from the 25th to the 75th percentile, with horizontal lines indicating the median value. The whiskers extend up to 1.5 times the size of the box if such observations exist. Observations (i.e., countries) that fall outside this range are shown as dots. Data from 2015 or 2014, depending on the latest observations available. Sources: IEA (2017); SE4ALL (2017a).

populations suffer more from energy poverty than their urban counterparts do. If access is available, the provision of modern energy is intermittent and rarely reliable. In the case of electricity, connection charges and market prices are often prohibitive. This is also often true for cooking, as transitioning toward more efficient fuels and less hazardous stoves is particularly expensive in rural areas.

Explaining Variation in Energy Poverty: The State of the Art

The considerable differences in energy poverty around the world raise the question of why there is such variation. While much has been said about the origins of energy poverty, it turns out that there is a gap in the literature. Although existing accounts link energy poverty to the availability of energy resources, economic development, and a country's geography, these accounts cannot explain variation across countries with broadly similar endowments. Institutional explanations that specifically tackle energy poverty as a policy problem include the lack of democratic governance, urban bias among rulers, and the weak design of national policies and institutions for energy access. These explanations are not satisfactory, however, because they cannot provide an overarching theory of energy poverty applicable to a large number of countries.

The simplest explanation for why some countries suffer from energy poverty is the availability of energy resources (Madureira 2008; Bhutto and Karim 2007). Countries that are very rich in energy resources rarely have difficulty granting their population access to electricity and modern cooking fuels. If energy resources are already widely available to countries, they do not have to buy expensive energy resources, such as oil or natural gas, from world markets. Besides providing an easy source of revenue, resource availability reduces the costs for energy provision considerably and makes eradicating energy poverty easier.

This simple argument can explain why electricity access is almost universal in countries like Saudi Arabia or Venezuela, which benefit from their enormous oil wealth. The same is true for Paraguay, a country that produces more than 80 percent of its electricity from hydroelectric facilities. The Itaipu Binacional dam along the Río Paraná at Paraguay's southeast border with Brazil has not only been the world's largest power plant for more than thirty years, but alone is basically sufficient to meet the country's entire electricity demand. In fact, revenues from electricity exports from this dam generate about a quarter of Paraguay's government revenues, accounting for almost 5 percent of its annual GDP.[13]

While the resource argument is appealing and intuitive, it has its limitations. Many countries in Western Europe could lift their populations out of energy poverty a long time ago despite limited access to indigenous energy resources (Fouquet 2008; Smil 2010; Schurr and Netscher, 1960). Chow, Kopp, and Portney (2003, 1528) emphasize this point by arguing that "simply possessing large fossil energy reserves is of questionable value to a country's development if there is no well-functioning and adequately equitable socioeconomic system enabling it to extract and deploy those energy resources for their full social benefit." As it turns out, there is no linear relationship between energy resources and energy poverty.

Consider the cases of Yemen or Nepal. Despite declining oil and gas reserves, Yemen is still well endowed in terms of natural resources, but its electrification rate remains below 50 percent (IEA 2017), a striking anomaly in the Middle East, where national electrification rates are usually above 95 percent. Lacking electricity access is often attributed to mismanagement in the public electricity utilities, which has raised barriers to badly needed infrastructure investments (El-Katiri and Fattouh 2011; Sovacool 2012b). Similarly, vast hydropower potential in Nepal remains unused (Pokharel 2001), which is troubling for simple, linear theories focusing on energy resources.

Another argument put forward to explain energy poverty is that a country's geography determines whether it can effectively address the problem. Large countries with inhospitable landscapes and large populations living in remote regions face a much bigger challenge of providing basic energy services than smaller and more densely populated states do. Niez (2010, 12) points out that rural areas that lack access to electricity can often be characterized by "a reasonable distance from national or regional electricity grids" and are mostly found to be "far from urban centers with a difficult terrain such as large rivers or jungles," or may "suffer harsh climatic conditions," all of which make grid extension hard to achieve. Since the provision of energy services generally benefits from centralized production due to economies of scale, geography is an important factor that clearly affects a country's energy poverty. Deichmann et al. (2011) and Kemmler (2007), for example, show that for electrification in sub-Saharan Africa and India, population densities and spatial distributions of households and villages are key to understanding energy poverty as vast distances and sparsely populated areas pose a challenge to conventional approaches of grid electrification. Building on this work, electrification in sub-Saharan countries seems to be qualitatively different from electrification in other developing countries because of the sheer size and scattered nature of rural populations (Onyeji, Bazilian, and Nussbaumer 2012). On a more microlevel, Kemmler (2006) points out that spatial heterogeneity in electrification rates across India is in part due to state boundaries and village structures.

All this is to say that geography and the distribution of energy consumers in space inform our understanding of determinants of energy poverty. Cost savings from large-scale, highly standardized, and centralized power production can easily evaporate in geographically vast countries because of high transmission and distribution costs, rendering nationwide energy provision problematic. While there is a strong case for the geographical determinants of eradicating energy poverty, geography is not fate. As Onyeji, Bazilian, and Nussbaumer (2012, 520) put it, geography does not remove the "requirement ... for clear political commitment and leadership with a strong focus on providing electricity access to the rural poor." Even favorable geography does not remove the need for infrastructural investments; conversely, a committed government can choose to invest in removing energy poverty despite a geographic handicap.

Brazil is an insightful example of the limitations of the geography argument because it has achieved high levels of rural electrification and almost universal access to modern cooking fuels despite being the fifth largest country worldwide (Pereira, Freitas, and da Silva 2010; Coelho and Goldemberg 2013; van Els, de Souza Vianna and Brasil Jr. 2012; Gómez and Silveira 2010; Jannuzzi and Goldemberg 2014). Contrasting with the argument about the role of spatial distances and energy poverty, electrification rates in Brazil are not only high in its urban centers of Rio de Janeiro and São Paulo, but also among the less than 2 percent of the Brazilian population living in the remote areas of the Amazonian region. This success in expanding energy access was achieved despite soaring costs for grid extension amounting up to US$4,000 per capita according to Niez (2010) and the establishment of an "automatic delivery system" (Lucon, Coelho, and Goldemberg 2004, 85) for LPG gas throughout the country. Independent of a municipality's location, the Brazilian government was highly committed to consolidating an effective infrastructure for energy services so as to provide electricity and clean cooking fuels to all households in Brazil. This hints at the importance of strong political will, as well as capable and efficient institutions, which are factors that are overlooked in geographical approaches.

Providing modern energy services is expensive, and the problem often is that "the financial resources required for doing so are in scarce supply" (Foley 1992, 145). This scarcity establishes an immediate link between low levels of income and energy poverty. Wealthy governments can afford investment into infrastructure, while poor governments cannot. The supply of energy services is capital intensive, which is why these services can only profitably be provided if the costs can be recouped by investors, which raises a considerable roadblock to rural electrification. Barnes (2011) emphasizes the high initial cost of grid extension, and Golumbeanu and Barnes (2013) demonstrate that access to electricity in rural sub-Saharan Africa may not improve much even when power lines are built and villages are connected to a national grid because households cannot afford to buy electricity. Prohibitively high connection charges, sometimes as high as a country's annual per capita income according to Bernard (2010), bar poor households from gaining access to electricity. Greenstone (2014) also identifies low income levels and high switching costs for electricity as main obstacles to rural electrification, which is echoed by Onyeji, Bazilian, and Nussbaumer (2012) in their econometric analysis across sub-Saharan states.

The other reason that wealth is relevant is that households with disposable income can pay for energy access. According to Foley (1992), a minimum amount of disposable income must be available to households before energy services become meaningful. No one is interested in electricity per se, but in the appliances one can run with it, such as lighting, mobile phone charging, fans, radios, or televisions. While subsidizing energy provision and offering, for example, graduated electricity tariffs may be a way forward to provide energy services to the most deprived and poorest households, systematically distorting price signals in markets biases incentives for private energy providers and puts yet another financial burden on government budgets. Once economies grow and societies develop, and incomes increase and households become richer, providing energy access becomes easier, thereby inducing a strong link between income and energy service supply. This is why energy poverty is much more prevalent among the least developed countries in sub-Saharan Africa and poor Southeast Asian states, but it also explains why these states find it especially difficult to address energy poverty effectively.

Increasing wealth is probably the most powerful single explanation for the eradication of energy poverty, yet it is not satisfactory in several respects. Indeed, Khandker, Barnes, and Samad (2012) note that if energy poverty were a straightforward consequence of income poverty, the entire concept of energy poverty would be unnecessary. There are poor countries that have made considerable progress in rural electrification and the provision of modern cooking fuels. In Senegal, we have already noted, the use of traditional biomass has decreased precipitously thanks to effective government policies. Ghana's rural electrification program began at a time of economic hardship and instability. In South Africa, the apartheid government's wealth and resources did not go toward rural development for the black poor. All of these examples underscore the importance of residual energy poverty once we account for wealth. A second reason to avoid conflating income and energy poverty is that attributing the eradication of energy poverty to wealth is analytically problematic. Because energy access produces economic benefits (Greenstone 2014), it could well be that the causal arrow points in the opposite direction. In sum, the strong correlation between economic wealth and energy access may not be a causal effect.

Energy resources, geography, and income all clearly exert an influence on how energy poverty can be successfully tackled by governments. Yet all these explanations overlook the importance of institutional factors. Barnes (2007) brings these factors to the forefront when he emphasizes how incentives and institutions influence the success of rural electrification. Even in a poor country, political leaders may be able to design policies, such as tailored tariffs, graduated prices, or flexible payment schedules, to bring energy services to their citizens. Carefully crafted policies have consistently been shown to contribute to successful rural electrification programs. Pritchett (2005) makes this point clear and highlights that policies need to be technically sound, administratively feasible, and politically robust—regardless of energy resources, geography, or income levels. This is essentially why careful planning and effective implementation, when paired with strong political commitment (Birol 2007), can make a difference in providing sustainable solutions to energy poverty. In designing these policies, the preferences of local populations need to be taken seriously to make sure that policies meet local needs (Barnes 2007; Niez 2010).

Broader institutional accounts from political science make us expect to see less energy poverty in democratic countries. For one, democratic institutions allow citizens to freely express their preferences and make these preferences heard by policymakers. Furthermore, politicians who want to stay in office depend on votes from their constituents, which incentivizes them to consider the electorate's needs (Przeworski 1991; Wittman 1995; Lake and Baum 2001; Adserà, Boix, and Payne 2003). Political competition in democratic systems emphasizes voter preferences more than autocratic politics, so one might expect energy poverty to be a lesser problem in democracies. High electrification rates in Mauritius, a definitive leader on the African continent, is a clear case in point here. A country where political accountability is high, institutions are capable, and corruption levels are low, Mauritius makes a good example for why democracies can be expected to perform well when it comes to providing comprehensive energy services (Congleton 1992; Neumayer 2002; Li and Reuveny 2006).

But democracy is no silver bullet. For example, rural electrification may not reach areas that are not politically pivotal (Crousillat, Hamilton, and Pedro 2010). Barnes, Khandker, and Samad (2011, 260) also describe cases in which local politicians distort decisions on grid extension and make biased decisions on pricing, bill collection, and disconnection policies on purely political grounds. Similarly, many reports suggest that subsidies are often paid to middle-class families that may not need these subsidies to get access to energy services but promise higher electoral rewards than do low-income families living in the countryside (Kojima, Bacon, and Trimble 2014). The Nigerian experience draws attention to the fact that democratization is insufficient as long as it does not go hand in hand with the implementation of regulatory agencies that can be characterized by high institutional capacity (Olukoj 2004). This is why an explanation of energy poverty that purely revolves around democratic electoral incentives may not be enough.

Yet another impediment to eradicating energy poverty in rural areas results from government bias toward urban constituencies (Lipton 1977; Bates 1981; Bezemer and Headey 2008). There are many reasons why this bias can be so persistent. Binswanger and Deininger (1997) make the point that agricultural production historically was characterized by an ample supply of cheap labor. However, the benefits from agricultural production in the form of cheap food prices were mostly reaped by the middle class and elites in cities, who felt no need for investment in agricultural production. These structural deficiencies got cemented in times of colonialism during which the governing elites resided in the colonies' urban centers, passing down this urban–rural bias even after many former colonies gained independence (Lipton 1977; Binswanger and Townsend 2000). The vision in many former colonies of a quick transformation into industrial production and service industries to foster economic growth aggravated the neglect of rural areas (Bezemer and Headey 2008). From a political perspective, rural households are disadvantaged not only structurally but also in terms of making their voices heard. Rural populations face much higher organization costs because they are typically far away from the political centers of power and are often isolated from each other. Forming interest groups and building effective political representation are much more difficult in rural areas, which leads to worse access to the political elite and pivotal decision makers (Bates 1981; Binswanger and Deininger 1997).

Urban bias is also related to domestic political institutions. Countries with democratic institutions are governed by the principle of electoral competition, which alleviates urban biases. This effect should be even stronger when rural populations are large and favored by electoral rules. Kale (2014a) attributes India's aggressive rural electrification push in some areas, such as Maharashtra, to the power of landowners who needed electricity to increase the productivity of their farming through irrigation and mechanization. Consider also the notion that for autocratic leaders, the best strategy to retain political power is to offer perks to their cronies in return for continuous political support (Olson 1993; Wintrobe 1998; Bueno de Mesquita et al. 2003). These political incentives worsen the urban bias and disenfranchise rural populations.

But urban bias is itself not enough to explain the lack of progress in energy poverty. India's push for rural electrification for agricultural reasons has enabled household electrification, but the provision of inexpensive or even free power has also driven state electricity boards to the brink of bankruptcy and, in doing so, impeded badly needed investments (Joseph 2010; Chatterjee 2012). The urban bias could even drive people to move into cities, reducing the need for costly investments into rural electrification. For example, while Brazil's geography has made rural electrification difficult, the country's large urban population has created a large revenue base for electric utilities.[14]

Finally, we can perceive the failure to adequately address energy poverty as an artifact of failed policies, spanning the entire range from initial steps toward privatization to more detailed policies about how to differentiate electricity fees, how to design subsidies for cost recovery, and how to structure modes of rural electrification (Bacon and Besant-Jones 2001; Kojima, Bacon, and Trimble 2014; Besant-Jones 2006; Vagliasindi and Besant-Jones 2013). Over the past three decades or so, many countries, both industrialized and developing ones, have seen power sector reforms that, albeit to varying degree, were intended to transform a state-dominated electricity sector into a competitive market (Victor and Heller 2007; Victor, House, and Joy 2005;

Rufín, Rangan, and Kumar 2003). As Eberhard (2004, 14) argues, there are usually three "broad drivers for power sector reform." First is a need to unbundle publicly owned enterprises in generation, transmission, and distribution of power, which lack accountability to private stakeholders. Second, continued electrification and energy access requires expansion of the grid and the energy infrastructure for which private investment is needed. Finally, power sector reform provides a window of opportunity for political leaders to redistribute rents by selling off public assets and reducing government debt. Besides privatization, several countries have attempted to move away from a top-down approach, instead mobilizing local stakeholders (Jewitt and Raman 2017). The notion that a bottom-up approach may be more successful is embraced by high-profile nongovernmental organizations (Practical Action 2016, 2017) and a theme that we explore in greater detail in our theory.

The standard textbook model (Victor and Heller 2007) of power sector reform involves allowing independent power production by private players; corporatizing state-owned enterprises; drafting legislation for electricity sector liberalization; the unbundling of generation, transmission, and distribution services; privatization; establishing wholesale markets; and allowing customers to choose their retail suppliers (Erdogdu 2011; Dubash and Rajan 2001). Some scholars, however, doubt it can supply modern energy services to the poor (Dubash 2003; Victor 2005; Goldemberg, Rovere, and Coelho 2004; Powell and Starks 2000), as market competition does not lead to the provision of services that are risky or unprofitable (Kojima, Bacon, and Trimble 2014). Victor (2005, 2), for example, concludes that "there is no simple relationship between power sector reform and energy services for the world's poor."

While this discussion focuses on power sector reform, a similar argument (which also emphasizes the importance of institutional and political contexts) is the idea that energy poverty can be addressed effectively if the implemented policies are designed well. The flip side of this claim is that inadequate policies, scarce financial resources, and weak institutions make policy failure very likely (Haanyika 2006; Barnes 2007; Chaurey, Ranganathan, and Mohanty 2004). Supplementing this, there is ample evidence from various regions around the world suggesting that the features of particular policies that are meant to ameliorate energy poverty do matter. Bambawale, D'Agostino, and Sovacool (2011) show the rampant increase in electrification in Laos in only about fifteen years was achieved by means of careful policy design. Keeping an eye on the commercial viability for electricity suppliers, offering flexible payment plans in the form of fee-for-service models, and knowing about the geographical and socioeconomic realities proved to be essential. Examples from Zambia (Haanyika, 2008) and Nepal (Yadoo and Cruickshank 2010) attest to this logic. The Nepal case study, for instance, examines the role of local cooperatives and finds that their systematic involvement in Nepal's rural electrification strategy proved immensely helpful.

Policy design is by no means inconsequential for the policy outcome, yet it remains an open question how approaches that explain successful provision of energy services as a function of policy design can inform our understanding of the conditions that bring about these good policies. Policies are never formulated in a vacuum, and their origins are themselves a major puzzle that any theory of energy poverty must address. Why do some governments invest in effective and sustainable policies? Why do others not? This is where our theory, which we present in the next chapter, can make an important contribution.

Energy Poverty and Sustainable Development

While improved energy access requires better policies at the national level, it also shapes sustainable development at the global level. Different solutions to the problem of energy poverty have different implications for global demand for fossil fuels, with ripple effects on the international community's ability to control climate change in the future. Energy poverty undermines economic development in poor countries, but some paths to better energy access are controversial because of their environmental side effects. The debate on energy poverty and sustainable development is important for understanding the policy challenges of today and tomorrow, especially in regard of new technologies such as solar power.

To begin, the development community worries about energy poverty because it slows economic growth (Shiu and Lam 2004). The lack of access to reliable energy reduces productivity, and the use of solid cooking fuels is an impediment to a healthy and productive life (Lam et al. 2012). At the turn of the millennium, the World Bank (2000, 1) commissioned a report on this topic, contending at the time that the "link between energy programs and poverty alleviation is less well understood—and more likely to provoke debate and soul-searching among energy specialists." These links are now better understood, with recent studies emphasizing how central energy access is related to poverty alleviation (Greenstone, 2014). Energy poverty is costly for the poor because the lack of access to modern energy requires a lot of investment in terms of time and effort into collecting inefficient sources of energy such as wood or dung. This is a problem in terms of both economic and human development because it prevents people from making better use of their skills (IEA 2011, 36). Energy poverty is also financially expensive. The *Economist* estimates that people who have to rely on kerosene instead of the grid to generate electricity pay about one hundred times more per kilowatt-hour compared to people who can rely on a modern electricity network.[15]

Recent studies, however, suggest that electrification on its own may have limited effects. Burlig and Preonas (2016) find that India's Rajiv Gandhi Grameen Vidyutikaran Yojana (RGGVY), a vast rural electrification program, did little to stimulate economic growth. Rwanda's Electricity Access Roll-Out Program similarly helped increase energy access but failed to markedly increase people's income (Lenz et al. 2017). Studies focusing on off-grid electrification generally paint a similar picture: technologies such as solar microgrids successfully reduce energy poverty and displace alternatives like kerosene, but there is scant evidence that they generate broader benefits such as new businesses or better educational attainment (Furukawa 2014; Grimm et al. 2016; Aklin et al. 2017; Kudo, Shonchoy, and Takahashi 2017). In a recent study, SE4ALL (2017b) reviewed the literature on the consequences of household off-grid electrification in twelve areas, including time spent studying and access to mobile phones. In only three of them was there "relatively strong" evidence that electrification benefits households.

The relationship between improved energy access and environmental sustainability is even more complicated. Though a number of more or less consistent definitions exist, the core idea of sustainable development is that current economic growth must not jeopardize future welfare (Solow 1993; Hopwood, Mellor, and O'Brien 2005). In other words, growth is sustainable to the extent that it does not undercut future growth. This may occur, for instance, if a society overuses a particular natural resource that is critical to its own well-being. The implication of the notion of sustainable development is that economic development may lead to abrupt reversals if people fail to keep an eye on the future. Short-term growth may plant the seeds for future hardship. Nowhere is this problem clearer than in the case of climate change, where the world's excessive reliance on fossil fuels raises global temperatures and threatens to cause climate disruption in the future (Stern 2006).

Overcoming energy poverty may necessitate the construction of new power plants. Often energy poverty is not caused by an inefficient use of existing energy resources but simply because of energy capacity constraints.[16] In other words, a lot of the infrastructure needed to generate enough electricity and clean cooking fuels to solve energy poverty still needs to be built. The critical question is how such a large amount of energy can be generated. If the experience of industrialized countries is any indication, the answer will be fossil fuels (Smil 2010). And therein lies the problem: coal, oil, and gas are all potent sources of climate change (Grubb 2006; IPCC 2013). If we have to choose between energy poverty and rapid climate change, we are truly choosing between two tragic outcomes.

Indeed, many prominent observers believe that there is a trade-off between reducing energy poverty and accelerating the pace of climate change. This idea is captured in a *Foreign Policy* column by Keith Johnson:

All those gleaming [new coal] plants, along with the inevitable controversies that surrounded their construction, underscore the fact that the world today faces two contradictory and interrelated challenges: While billions of poor people in the developing world need a lot more energy to pull them out of poverty and drive economic development, improve life expectancies, and bolster human health, the world also faces a looming and possibly existential threat from climate change.[17]

Thus, helping billions of energy-poor people may require the construction of power stations running on fossil fuels. And as individuals escape energy poverty, their growing income may create more demand for energy-intense goods, further increasing pressure for more power (Wolfram, Shelef, and Gertler 2012). From this perspective, any policy devoted to reducing greenhouse gas emissions is bound to have detrimental effects on policies designed to tackle energy poverty. For example, countries may need to increase cooking fuel prices in order to reduce their carbon dioxide emissions, but that would make modern cooking technologies such as LPG too expensive for poor people (Cameron et al. 2016).

The energy poverty-climate nexus led to vocal debates about the ethics and politics of this trade-off. Commenting on a report written for the iconoclastic Breakthrough Institute, Daniel Sarewitz, a professor at Arizona State University, and his colleagues claim that "climate change can't be solved on the backs of the world's poorest people."[18] Similarly, Charles Kenny, a fellow at the Center for Global Development, penned a column titled "Poor Countries Shouldn't Sacrifice Growth to Fight Climate Change."[19] According to this view, countries worried about climate change should not impose their concerns on countries that are desperate for economic growth. This, of course, does not mean that the proponents of this position are not worried about climate change. A noted proponent of energy growth, Bill Gates, contends that "instead of putting constraints on poor countries that will hold back their ability to fight poverty, we should be investing dramatically more money in R&D to make fossil fuels cleaner and make clean energy cheaper than any fossil fuel."[20] However, the question remains: Should we prioritize climate change mitigation or the eradication of energy poverty?

The emphasis on trade-offs has been challenged on three grounds. First, some reject the notion of a trade-off altogether. Research suggests that reducing energy poverty may have negligible consequences on greenhouse gas emissions. According to Pachauri (2014, 1073), "recent studies that have assessed the emissions implications of eradicating energy poverty globally or achieving universal modern energy access for cooking and electricity use in homes ... conclude that these efforts would contribute only marginally to greenhouse gas emissions over the next decades." For instance, Chakravarty and Tavoni (2013, S67) argue that "an encompassing energy poverty eradication policy to be met by 2030 would increase global final energy consumption by about 7% (roughly 20 EJ). The same quantity of energy could be saved by reducing by 15% energy consumption of individuals with standards above current European levels." According to these views, the trade-off suggested simply does not exist; energy poverty can be solved with little effect on the global climate.

Second, many commentators believe that renewable energy sources, which generally qualify as clean, can contribute to solving energy poverty. According to this perspective, energy poverty can be solved with energy resources that have little if any effect on the planet's climate. In particular, renewables may prove to be a sustainable energy source. Some scholars refer to such a process as "leapfrogging": instead of going through the polluting stages of their development, emerging countries can immediately adopt clean energy technologies (Zhang, 2014). Alstone, Gershenson, and Kammen (2015, 313) argue for a paradigm shift in energy policy, stating that

with a foundation of super-efficiency and carbon-free generation, supported by new ICT [Information and Communication Technology] connectivity and applications, expanding access through decentralized power systems could have radically different climate and equity impacts from the incumbent system, challenging the conventional knowledge held by some that one must choose between progress on energy access or climate.

This perspective is embraced by recent international initiatives tackling energy poverty. The United Nations devoted a special initiative, SE4ALL (Sustainable Energy for All), to this cause. SE4ALL provides a platform for policymakers, NGOs, and private sector actors to interact (SE4ALL 2011). In recent years, SE4ALL began to use its means to provide more direct help. For instance, SE4ALL intends to provide technical assistance on renewable energy and energy efficiency policies in the Indian state of Gujarat.[21] Earlier, in 2009, the International Renewable Energy Agency, a new organization, was founded to support countries "in their transition to a sustainable energy future."[22] Similarly, the IEA followed and sponsored the Energy+ initiative, which "aims to increase access to energy and decrease or avoid greenhouse-gas emissions by supporting efforts to scale up investments in renewable energy and energy efficiency" (IEA 2011, 38). These international initiatives all seek to facilitate a sustainable reduction of energy poverty, and all rest on the feasibility of leapfrogging.

Finally, some observers believe that the benefits of fossil fuels are overstated (ODI 2016). From a technical angle, energy poverty is caused by the absence of numerous interlinked infrastructures. For instance, electricity requires a power station, a transmission system, and connection devices. In the absence of a transmission network, new power stations are useless. A country may build hundreds of coal plants, but these will not shift energy poverty at all if the remaining infrastructures lag behind. In contrast, clean energy off-grid solutions such as solar microgrids may offer much better prospects in this regard. Because these solutions are flexible and require a much lighter infrastructure, they can be better adapted for rural communities. Offering a rebuttal to coal advocates, Joshua Hill observes in *CleanTechnica* that "without an energy grid, the energy generated by coal would have no way of reaching its intended recipients, requiring a massive outlay to build the necessary infrastructure to reach the very people renewables apparently 'can't.' ... Compare on-grid costs to mini- and off-grid costs for rural development of renewable energy, and there are still investment costs, but they are inherently cheaper and more effective."[23]

These debates also revolve around our idea of energy access. The environmental organization Sierra Club, for example, defines access to energy as "lighting, television, fan, mobile phone charging, and radio ... at least two hours of daytime agro-processing to support livelihood generation" (Craine, Mills, and Guay 2014). Focusing on electrification, this definition of energy access is arguably a minimal standard, and many of the arguments against renewables are based on the notion that much larger loads of power are necessary for modernization of developing countries. If we define a low threshold for energy access, the focus is on the provision of basic services. The downside of this approach is that it raises difficult questions about equality: Why should developing countries strive to guarantee only basic energy access when virtually everybody in the United States uses at least an order of magnitude more energy at home?[24]

The debates have concrete implications for government and international organization policy. To see this, consider how the Overseas Private Investment Corporation (OPIC) of the United States, which provides capital for private investment that promotes economic development and advances American interests in foreign policy, has a policy that discourages funding for fossil fuels. Writing for the Center for Global Development, Moss and Leo (2014) claim,

There has been a general bias toward using OPIC to invest principally in solar, wind, and other low-emissions energy projects as part of the administration's effort to promote clean energy technology. ... We estimate that more than 60 million additional people in poor nations could gain access to electricity if OPIC is allowed to invest in natural gas projects, not just renewables.

While their methodology can be questioned on many grounds, including the decreasing costs of renewable energy, the fact remains that any policy on development funding or lending that excludes fossil fuels does reduce the number of options available for investments to reduce poverty. Although the exact cost of these restrictions in terms of energy poverty is unclear, the direction of their effect on energy poverty eradication is clearly negative.

While this book does not directly engage these normative arguments, we highlight that our goal here is to explain variation in the success and failure of government efforts to reduce energy poverty. From this perspective, success does not depend on whether the mitigation of energy poverty has caused an increase in the use of fossil fuels. While we are concerned about the effects of climate change on human societies and ecosystems, we do not believe these concerns are critical to understanding government policy. As we shall see, governments of developing countries invest in fossil fuels if they see concrete political or economic goals from doing so, even at the expense of environmental degradation. Progress in the use of renewable energy does promise to mitigate the trade-off between climate change and energy poverty in the coming years, an issue that we return to in the concluding chapter.

3   The Political Economy of Energy Poverty

The prevalence of energy poverty among developing countries today masks substantial variation in their trajectories. Some countries have progressed in their efforts to reduce energy poverty in leaps and bounds over the past two decades. Others have achieved little. Furthermore, there is much more progress in rural electrification than in the provision of modern cooking fuels. How can we explain this variation? How can we explain the varying energy access trajectories of countries that had achieved virtually no electrification or modern fuel access for cooking before the beginning of World War II—that is, most of Asia, Africa, and Latin America?[1]

As we showed in chapter 2, previous explanations do not provide a comprehensive and structured account of the sources of success and failure in the eradication of energy poverty. They are often based on individual case studies and downplay political considerations despite the clear evidence of the importance of government policy. In light of these limitations, we present a broadly applicable account of the political economy of energy poverty. Emphasizing political considerations and government behavior, our goal is to develop a systematic, empirically falsifiable theory of energy poverty. Drawing on the first principles of political economy, we can shed new light on the why and when of progress in mitigating this policy issue. The scope of the theory is not limited to rural electrification or access to modern cooking fuels alone, but rather allows us to apply a unified analytical framework to these two faces of energy access. Since energy poverty is predominantly a rural issue, however, we mostly focus on the challenge of providing energy access to villagers.

Our theory emphasizes energy access at the expense of energy production, but this does not mean that production is irrelevant. If public electric utilities are governed by policies that enable them to cover their generation costs and produce profits, then these profits can be invested to improve rural electrification. This may trigger a virtuous cycle, whereby any electricity connections created will be more useful to the people who gain energy access, as the availability and reliability of electricity—daily hours and the frequency of outages—will be higher than in the case of a bankrupt electric utility. Similarly, solving the problem of cooking fuels is easier if the country has access to domestic natural gas resources and the conversion of those resources into LPG is cost-effective.

Our theory is based on a series of assumptions about government behavior and other factors, spelled out in the next section. For now, it suffices to say that our fundamental premise is that the national government is the primary actor in efforts to reduce energy poverty. Since the reduction of energy poverty requires massive investments in infrastructure, the government must either use public funds or, at the very least, create a policy environment that encourages and supports private investment. Due to the scale and capital intensity of eradicating energy poverty at the national level, a completely market-driven solution without any government participation or support is not realistic. Indeed, the primary beneficiaries of energy poverty alleviation are marginalized members of the society, so a market-driven solution might exclude the poorest and most deprived households. This logic stands in stark contrast to, say, the construction of road infrastructure, which directly serves industrial and business interests. The government itself is not considered enlightened, altruistic, or omnipotent. In a political economy account, the government's behavior reflects, along with intrinsic preferences and ideology, concerns with political survival and constituency pressure (Olson 1993; Bueno de Mesquita et al. 2003; Baccini and Urpelainen 2014). These concerns and pressures depend on the nature of political institutions, the prevailing socioeconomic conditions, and the availability of resources.

The basic logic of our theory is summarized in figure 3.1. We focus on the ability of policy to reduce energy poverty, holding factors such as income and geography constant. Possible outcomes vary from failure—no progress on eradicating energy poverty through policy—to partial and complete success. In the case of partial success, energy poverty is eradicated but possibly slowly and unevenly across the country, with lingering concerns about the sustainability of progress. Ghana is a good example. The country made considerable progress over the past two decades as it gradually transitioned

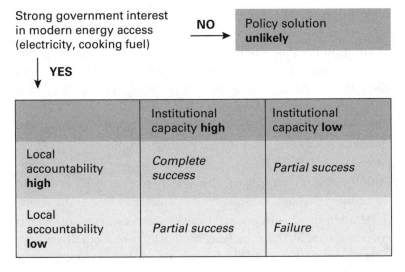

Figure 3.1
The building blocks of the theory.

from an autocracy to a democracy. In recent years, however, blackouts due to power shortages and poor transmission lines threaten to inflict a bitter setback.[2] In the case of complete success, energy poverty is eradicated rapidly and evenly in a cost-effective and sustained fashion. In other words, complete success means the government is, controlling for structural factors such as wealth and geography, seizing on the vast majority of available opportunities to reduce energy poverty through policy. Complete success in policy does not require 100 percent rural electrification or access to clean cooking fuels, but it does require fast and sustained progress. We document such cases, including China, in chapter 5.

If the national government has little interest (interchangeably, incentive) in eradicating energy poverty, a policy solution to the problem is unlikely. Energy poverty is a political problem, and solving it requires forceful political action. Next, we go beyond stating this intuitive fact by characterizing the conditions under which a government's interest in solving the problem results in policy success. Here, we emphasize the effects of two factors: local accountability and institutional capacity. By local accountability, we refer to the ability of the rural victims of energy poverty to voice their concerns and punish politicians for failing to act; institutional capacity refers to the government's access to effective implementing agencies—such as electric utilities, rural electrification agencies, and public LPG distribution networks—for improving energy access. Both factors have an independent positive effect on the probability of policy success. When neither is present, the government's interest is not sufficient for progress. The condition most conducive to reduced energy poverty combines high levels of local accountability and capable institutions.

Within reasonable bounds, our argument can be applied at the subnational level. So far, we have focused our discussion almost exclusively on the national government's interest and considered capacity and accountability mostly with respect to large-scale energy programs. Yet as we will see in subsequent chapters, a similar story emerges if we look at the subnational level. For instance, institutional capacity varies tremendously within large countries such as India. In Bangladesh, different institutions have been tasked over the years with improving electrification rates. Some of them, such as the Rural Electrification Board, have scored high on both capacity and accountability, whereas others have failed.

The next section discusses our basic assumptions about the economics and politics of energy poverty in some depth. We then draw on these assumptions to explain why governments are sometimes—but not always—willing and able to tackle energy poverty. Then, we consider the effects of the introduction of new technologies, such as off-grid solar power and efficient cookstoves, on the political economy of energy poverty. The final section applies these general principles to rural electrification and access to clean cooking fuels, our two substantive domains of interest, contrasting the prospects of government across these two domains.

Building Blocks: Economics and Politics of Energy Poverty

To explain and understand energy poverty, we need to make some analytical assumptions about economics and politics. Recall that by energy poverty, we refer to a situation in which households are unable to meet basic needs with respect to two fundamental facets of energy access: electricity and modern cooking technologies. There is a wide consensus among scholars about the multidimensional nature of the problem. Energy poverty is a complex social phenomenon that reflects both underlying economic constraints and the inability of government policy to relax them. To simplify the problem and gain analytical leverage, we focus on the national government as the primary actor.

An implication of our focus is that energy poverty may, or may not, pose a problem to the government. The government's interest in acting depends largely on whether it sees a gain in helping households meet their basic energy needs. In turn, if energy poverty is a problem from the government's perspective, then its ability to make progress toward solving it depends on a variety of structural factors. Even the most enthusiastic government may fail to act if the setting of policy formulation raises high barriers to effective action.

Our focus on the rural poor as the primary victims of energy poverty is a practical consideration, but it has implications for the focus of the analysis. The rural nature of energy poverty puts a heavy emphasis on infrastructure. Although there is widespread variation in the cost of connecting different households and communities to the electric grid or organizing a reliable supply of modern cooking fuel, rural infrastructure is generally much more expensive on a per capita basis than urban infrastructure is. In urban areas, energy poverty is a problem that to some extent is being solved as a by-product of the provision of good service to powerful business interests and the urban middle class.

Without any policy support, the ability of the rural poor to solve the problem of energy access is limited. They may, for example, purchase biomass cookstoves or use municipal resources to construct small power generation facilities, but these measures are usually not an effective solution to the problem because rural communities rarely have the resources to go beyond basic achievements. In practice, as we shall see in the case studies, effective local action is often supported by government agencies and policies at higher levels. Tackling energy poverty is thus beyond the means of individual citizens and communities, so government must play a key role. Indeed, even a market-based solution often requires policies such as subsidies and tax breaks to ensure private actors can invest profitably.

Consider the case of rural electrification. Setting aside the issue of power generation itself, rural electrification requires the construction of a transmission and distribution network. If the customers are poor and therefore the revenue derived from their consumption is low, their power consumption is not sufficient to cover the high capital costs of grid extension. Thus, public investments have an important role to play.

In the case of cooking, the challenges are similar. If the government wants to distribute LPG or efficient biomass cookstoves to rural households, it incurs many expenses. If poor households cannot afford these cookstoves, the government needs to provide either subsidies or credit. In the case of LPG, if suppliers are unwilling or unable to cover the cost of distributing the fuel, the government has to support the creation of a distribution network. The LPG fuel itself may require subsidies if it is not affordable to poor households. Efficient biomass cookstoves require a maintenance solution, and private actors hesitate to commit to offering good service in rural communities to poor households over long periods of time. If the government entices private actors to solve the maintenance problem, the problem of oversight remains.

Drawing on a large body of literature in political economy (Wintrobe 1998; Lake and Baum 2001; Bueno de Mesquita et al. 2003; Baccini and Urpelainen 2014), we next assume that the government's primary consideration is political survival. While all governments have intrinsic preferences and ideological views, their common concern is to remain in power. If a government loses power, it can neither pursue policy goals nor use political power to secure private benefits. Except in rare circumstances, governments continuously worry about threats to their survival. These threats include votes of confidence, upcoming elections, revolutions, and coups. This does not mean that ideology is absent from the policymaker's calculus; rather, ideology is often conflated with survival needs. For instance, a government that seeks to alleviate poverty may do so for ideological reasons and to remain in power. Both autocratic and democratic governments must worry about their political survival, but the nature of the threats they face can be different.

The primacy of political survival implies that a government's policies reflect the preferences of politically powerful constituencies: groups of people in the society who have shared interests in regard to policies that can eradicate energy poverty. Constituencies are considered politically powerful if securing and retaining their support is important for the political survival of the government. In the case of energy poverty, the question is whether societal interests that expect benefits from mitigating energy poverty have enough political clout to sway government decisions. In practice, the core constituency is the population of the rural energy-poor themselves. If these people have political power, through the ballot or other means, the government has an interest in catering to their needs. Many studies of political accountability over the past few decades suggest that governments set policies to ensure that their performance is adequate to retain the support of politically pivotal constituencies (Ferejohn 1986; Bardhan and Mookherjee 2006a; Fearon 2011). Although the logic of accountability varies widely across contexts, no political leader can survive without some constituency support.

To strengthen its prospects of political survival, the government can allocate benefits to politically powerful constituencies in two primary ways. One is to promote economic growth. If the national economy grows faster, more resources are available to everyone. The other is to redistribute available resources. If the size of the pie remains the same or even if it shrinks, the government can cater to specific constituencies by taking resources from one group and giving them to another. As we shall see, both mechanisms play a role in government efforts to mitigate energy poverty.

Because political scientists often emphasize the institutional setting as an explanation for the government's interest in providing benefits to different groups, it is useful to begin with the role of democratic versus autocratic political institutions. In the case of energy poverty, the question is whether the rural energy-poor are an influential constituency with political clout. Consider, first, the case of an autocracy. Beginning with Lipton (1977) and Bates (1981), political science research has documented an urban bias, especially under authoritarian regimes. Since rural dwellers have less capacity to protest and riot to overthrow an authoritarian regime, a dictator's best strategy is to concentrate resources to placate people living in urban areas, especially the capital. In contrast, democratic regimes depend on rural votes for their electoral success. Since rural populations are large in most developing countries and often politically influential even when small due to institutional "malapportionment" (Samuels and Snyder 2001) of votes, democratic regimes generally have stronger incentives to implement policies that favor the poor. Because democratic politics favors the rich and the powerful relatively less than autocratic politics does, we would expect democratic political institutions to be more conducive to eradicating energy poverty than autocratic political institutions would.

However, the democracy-autocracy difference alone is not enough to predict government action to combat energy poverty because of the considerable variation in the logic of voting among democracies. In some countries, voting is mostly based on clientelism and patronage (Kitschelt 2000; Stokes 2005; Keefer 2007; Thachil 2011), meaning that politicians offer private benefits, such as consumer goods or cash, in exchange for votes. In such settings, democracy need not lead to infrastructural investments to reduce energy poverty. If politicians campaign for votes based on a quid pro quo, major infrastructural investments should not be on the horizon. If citizen-politician linkages are based on direct exchanges of votes for private goods, democracy need not result in effective public investments into the eradication of energy poverty. In other countries, voters support parties that implement policies that solve their problems in a "programmatic" fashion (Hagopian, Gervasoni, and Moraes 2009). Under a programmatic political logic, voters reward political parties for effective policy; the parties then tend to support an agenda that is more favorable to the provision of public goods. In these settings, governments have a stronger interest in removing energy poverty than under conditions of clientelism. If the energy-poor refuse to reward politicians for handouts, they can pool their electoral power and demand policies that effectively eradicate energy poverty.

These direct interests in removing energy poverty are important, but the government may also act to reduce energy poverty for other reasons. Often measures that effectively reduce energy poverty also produce other benefits. Consider, for example, the case of rural electrification. Improved electrification of the countryside also improves agricultural productivity by enabling irrigation and mechanized farming, benefiting farmers and the national economy (Cabraal, Barnes, and Agarwal 2005; Brass et al. 2012). A government that does not care about the energy-poor may nonetheless solve their problems as a strategy of rural development.

Even if the government has some interest, whether direct or indirect, in eradicating energy poverty, it has to consider the cost of acting. At any given time, the government has many problems to address. If the cost of acting to reduce energy poverty is high, then action on this front compromises the government's ability to address other pressing concerns. As to the factors that determine the cost of acting to reduce energy poverty, we place a premium on considerations of institutional capacity, that is, the government's capacity to implement effective policies at a low cost. A high institutional capacity requires a competent and organized bureaucracy, enough resources for implementation, and the ability to commit to plans over long periods of time. Of course, institutional capacity itself requires government efforts to build it over time. For our analysis, we accept institutional capacity as a given, noting that capacity is built very slowly. Since institutional capacity is a slowly changing factor, it is possible to assume that the foundations of institutional capacity were laid much earlier than the foundations of energy poverty alleviation at any given time.

An additional clarification regarding the origins of capacity is needed. For institutional capacity in the energy sector, it is useful to make a clear distinction between domestic and international capacity. The foundation of institutional capacity is the domestic administrative apparatus, but historically both multilateral and bilateral agencies have offered generous support to the eradication of energy poverty, especially for electricity generation (Hausman, Neufeld, and Schreiber 2014). Rural electrification projects are a staple for the World Bank, for instance.

Our final assumption concerns the stickiness of progress. We believe that successful efforts to eradicate energy poverty are somewhat sticky. If a government achieves great success in improving access to clean cooking fuels, for example, these gains are not easily lost: it is easier to keep them than to gain ground. However, dramatic changes for the worse in the conditions for energy poverty mitigation do result in backsliding in the long run. For example, the electricity grid requires maintenance. If a government manages to extend the grid to remote rural villages but a subsequent government has little interest in continuing to provide electricity access to these villages, deelectrification occurs over time.[3]

When Do Governments Tackle Energy Poverty?

We are now ready to answer the question of when decisive and effective government action on energy poverty may be expected. Based on the assumptions we have made, the government's decision depends on the relative costs and benefits of formulating and implementing new policies. The analytical task at hand is the identification of the conditions under which the benefits exceed the costs so that action is warranted.

Under conditions of rapid economic growth, especially in rural areas, action on energy poverty is an obvious choice. Economic expansion provides both the government and the society with resources that can be used to gain access to modern energy. The expansion of private resources means that households have more disposable income to pay for services such as a household electricity connection or LPG fuel, while economic growth allows the government to generate public resources through taxation. These resources relax the government's budget constraints and allow it to invest in energy poverty eradication. As we have already seen, the association between economic development and energy access is strong and robust. While energy access contributes to economic development, the growth of the national economy also improves energy access by providing resources.

The role of the economy and resources can be readily seen in the case of the Middle East. As we saw in the previous chapter, this area of the world has largely managed to escape energy poverty. This success reflects exceptionally favorable economic conditions. After the 1973 and 1979 oil crises, most Middle Eastern countries reaped huge windfall profits from oil and gas. These resources allowed the local administrations to invest large amounts of money into various public infrastructure projects, including electrification and cooking fuel provision. In such favorable circumstances, progress in removing energy poverty is a matter of course, and nuanced distinctions of government interest, local accountability, and institutional capacity are secondary considerations. Unfortunately, such times of rapid economic growth are the exception, not the rule, in the developing world. In most years, the economy of any given developing country is either stagnating or growing at a relatively low rate. While such growth can, and has, made a massive difference over time in economic development, it does not mean that the eradication of energy poverty is an obvious choice. Resource constraints remain stringent, and the government must decide how to best use the limited resources that economic growth is making available. In this typical condition of resource scarcity, the government's resource allocation presents a vexing problem. Different political constituencies call for more resources, and a distributional conflict ensues. The energy-poor are but one constituency among many to demand the improvement of their lot.

What is worse, the eradication of energy poverty requires large infrastructural investments that carry a large price tag. If the government is to reduce energy poverty, it must create policies and programs that improve the energy infrastructure in the rural areas. Since this approach requires large capital investments, it is, at least in the short run, a clear drain on the government's resources. The constituencies whose needs go unmet complain and may even mobilize to support the political opposition.

The resources needed to successfully complete rural electrification are not trivial. To take but one example, in the mid-1990s, Morocco undertook the Programme d'électrification rurale global (PERG, or Global Rural Electrification Program). At the time, rural electrification rates hovered around 20 percent, but the program turned out to be extremely successful, leading to a rate of over 97 percent in 2009.[4] This success, however, was not free of cost. PERG is estimated to have cost over US$2 billion, and this calculus ignores spending on earlier programs that achieved little. In 1979, the International Bank for Reconstruction and Development (IBRD) provided a loan of US$42 million (at 7.9 percent), or about US$135 million in 2014 dollars.[5] Ten years later, the IBRD made a second loan of US$114 million—about US$200 million in 2014 dollars—at the IBRD's variable rate.[6] Fortunately, Morocco could rely on a network of outside funders for contributions. Institutions such as the French Development Agency, the Japanese Bank for International Cooperation, and the Islamic Development Bank all supported its effort (Islamic Development Bank 2013). Furthermore, energy corporations such as Total and Elf, both from France, offered technical support. Without such support, the government would have been forced to make painful trade-offs.

In normal times, large capital investments in rural electrification are not an obvious choice for any government, regardless of regime type. In the autocratic setting, urban bias undermines incentives to eradicate rural energy poverty. Urban protests are a more important threat to the government's political survival than mobilization in the countryside, and so the government's priority is addressing the concerns of urban constituencies. In democracies, the rural vote creates incentives to act on energy poverty, yet the specter of political opposition to capital investment is also present. Various interest groups compete for the government's attention in a process of political mobilization and coalition formation. Since the rural energy-poor are a marginalized segment of the community, their mobilization capacity and resources available to support it are often limited. Moreover, their deprivation means that their collective action is impeded by the temptation to support voters in exchange for particularistic benefits in a clientelist fashion. Indeed, direct and immediate benefits in the form of private goods are understandably lucrative relative to the uncertain promise of infrastructural improvement in the future.

Given these serious challenges, the government's interest in tackling energy poverty must stem from a strong interest in available benefits. Since there is always going to be some opposition to major infrastructural investments, the government's willingness to act depends on the expectation of resources that facilitate political survival. These resources can be direct or indirect.

The direct benefits of eradicating energy poverty are derived from increased political support among the energy-poor and their allies. In democracies, such political support means more votes in the next election. In autocracies, the energy-poor can be part of the autocrat's political support base. Despite the urban bias, riots and protests in the countryside are hardly in the interest of the autocrat, especially if they threaten agricultural productivity and the government's ability to supply food and other agricultural goods to urban dwellers. Therefore, the political clout of the energy-poor is a central determinant of government interest. Where the government's political survival depends on the support of the energy-poor, government interest is high; where the energy-poor have no political power, the government can ignore their needs without paying a cost. While democratic political institutions do not guarantee government interest and autocracy does not guarantee lack thereof, the logic of electoral competition means that democratic governments are more likely to be interested in the eradication of energy poverty.[7]

Recent work has shown that electricity plays an important role in electoral politics (Min and Golden 2014; Min 2015). Drawing on the Indian experience, these studies show that politicians in developing countries use electricity supply strategically to increase their vote share. Even in electrified areas, government officials can increase the supply of electricity to voters in election years, and often for free. Importantly, though, the relationship between this strategy and the eradication of energy poverty is unclear. As Min and Golden (2014, 624) note concerning transmission and distribution losses in Uttar Pradesh,

Although some line loss is unavoidable, the results of our data analysis suggest that much line loss is not a function of technical features of electricity provision but rather an outcome of processes that benefit politicians and some consumers in the short-run despite detrimental consequences to the state and its citizens in the long-run.

The opportunistic increase of electricity supply in election years hurts the finances of electric utilities, thus hurting the progress of rural energy development in the long run.

Besides these conceptual difficulties, the measurement of the political clout of the energy-poor presents notable empirical challenges. We deal with these difficulties in the empirical chapters; an illustration suffices at this stage. One approach to the measurement issue is to take advantage of constitutional reforms. For instance, democratization generally shifts power from a small ruling elite to a broader electorate (Bueno de Mesquita et al. 2003). More generally, we can identify whether a constitutional reform increases or decreases the importance of the energy-poor.

South Africa is an instructive case. Following the introduction of apartheid and a series of constitutional amendments in the 1950s, black South Africans were legally prevented from voting (Posel 1991). The ruling white elites controlled most levers of power and were significantly richer than the black population. Incidentally, they also benefited from much higher electrification rates (Bekker et al. 2008). Beyond the racial biases of lawmakers, the institution itself guaranteed a lack of political weight to the poorest segments of the population. The end of apartheid in the early 1990s marked a constitutional shift, empowering a vast number of previously disenfranchised—and mostly energy-poor—individuals. While the black majority could safely be ignored under the previous regimes, the new constitutional regime ensured that their political clout had changed. As we show in chapter 5, these institutional reforms coincided with a radical improvement of energy poverty (Bekker et al. 2008), but for now, it suffices to say that we leverage factors such as institutional reforms to measure the political strength of the energy-poor. The indirect benefits of eradicating energy poverty are only tangentially related to improving energy access, yet their importance for understanding government efforts to eradicate energy poverty is critical. These benefits can be reaped through the implementation of policies that also reduce energy poverty. Since the eradication of energy poverty is ultimately about improving rural infrastructure, policies that contribute to progress toward this goal can also contribute to rural development in many other ways.

The most important indirect benefit of measures to reduce energy poverty is greater rural productivity (Cabraal, Barnes, and Agarwal 2005; Cook et al. 2005; Dinkelman 2011; World Bank 2008). In modern agriculture and other rural businesses, access to energy is critical. Electricity access enables mechanization of the work, leading to greater productivity and reducing the hardship of farm labor. Power can be used to move things around and process raw materials. Liquid fuels can be used to power tractors, automobiles, harvesters, and other agricultural equipment. As rural productivity improves, less labor is required to produce a certain amount of food, which in turn frees labor for industry and services. In this sense, rural energy modernization can contribute to economic development at the national level. While this growth creates resources for the eradication of energy poverty, the causal arrow also points in the other direction: measures that eradicate energy poverty also contribute to economic growth.

Consider, for example, the case of irrigation. In the absence of modern irrigation technologies, harvests depend on benign weather conditions. In India, for instance, good conditions mean sufficient monsoon rainfall (Sekhri 2011, 33). Therefore, relying on good weather induces a high degree of variance in agricultural output. To the extent local geological conditions permit, farmers can avoid such uncertainty by drawing on groundwater to generate a more reliable and readily accessible source of water. Indeed, groundwater irrigation has been credited as a key factor in reducing the vulnerability of India's agriculture to climatic shocks (Repetto 1994; Sekhri 2011). However, using groundwater requires the construction of wells and the availability of pumps to extract the water. Pumps do not run without energy; in India, most run with electricity, though many use diesel (Shah et al. 2004). Unsurprisingly, then, agricultural output has become dependent on the reliable supply of affordable electricity to farmers in India.

The provision of irrigation also generates political benefits to the often influential agricultural lobby. The rapid deployment of wells is particularly valuable for wealthier farmers because pumps and the construction of wells are expensive (Sekhri 2011). Historically, only farmers with enough capital were in a position to take advantage of these new technologies (Repetto 1994, 42). In turn, this situation had two positive consequences from the point of view of policymakers. First, modern irrigation increased agricultural output, which put downward pressure on prices. This ensured that food prices would remain low and urban dwellers satisfied. Second, by favoring the most important farmers, the deployment of modern irrigation techniques was politically self-reinforcing. Instead of uniting the rural world against low agricultural prices, rich farmers had an incentive to preserve the status quo (Varshney 1993). These strong interests may also explain why electricity for agricultural purposes remains highly subsidized in many parts of India (Kimmich 2013b). A state such as Andhra Pradesh regularly spends more than 1 percent of its yearly budget on electricity subsidies for agriculture (Kimmich 2013a).

The network nature of the electric grid (Hughes 1993) amplifies the importance of these indirect benefits. If an economically or politically motivated government extends the electric grid to offer electricity for irrigation (Kale 2014a), this grid extension reduces the physical distance between a distribution point and the rural households in the area. Although this form of grid extension for purposes of irrigation, known as pumpset electrification, does not offer electricity for rural household use, it reduces the distance, and thus the cost, of household electrification in rural communities. As such, the very nature of the electric grid creates a link between productive and household electrification of rural areas.

The provision of modern cooking fuels can also have productive benefits. First, modern cooking fuels are less damaging to human health. The combustion of biomass increases the likelihood of illnesses such as lung diseases, asthma, or tuberculosis (Mishra, Retherford, and Smith 1999; Schei et al. 2004). According to the most recent estimates, cooking with solid fuels leads to the death of about 3.5 million people every year (Lim et al. 2013). This number does not reflect the years spent living with debilitating illnesses from biomass combustion. From an economic perspective, solid fuels have a detrimental effect on productivity by reducing the supply and productivity of labor. Second, cooking with solid fuels is a drain on time. Solid fuels are not very efficient and must be collected in large quantities on a regular basis (Cabraal, Barnes, and Agarwal 2005, 124). This can be time-consuming and often disproportionately affects children and women (Foell et al. 2011). Cabraal, Barnes, and Agarwal (2005, table 3) report estimates of the time spent collecting fuels from various developing countries in the 1980s. They find that these estimates typically range from three to four hours per day. Moreover, cooking itself takes longer with traditional fuels. A study from Bangladesh suggests that a median-sized rural household spends a bit less than four hours every day cooking (Miah, Al Rashid, and Shin 2009 table 2). The productivity loss is therefore not trivial. It is thus clear that modern cooking fuels have serious implications because they affect people's health and the time individuals have to engage in other activities.

Measures to increase rural productivity do not automatically eradicate energy poverty. If the government has no interest in energy poverty eradication, then productive use of energy may not improve the lot of the energy-poor to any significant measure. Consider again the case of irrigation. Even if the government draws distribution lines from the main transmission lines to electric water pumps, this does not automatically mean that households in the communities around the water pumps gain access. The ability of the energy-poor to benefit from their proximity to distribution equipment for irrigation depends on the cost of household connections, the monthly expense of basic electricity access, and so on. While productive use of electricity often reduces the cost of connecting poor households to the grid, it does not guarantee its improvement without additional government action.

The productive benefits of eradicating energy poverty are of interest to both autocracies and democracies. With the exception of the least capable, predatory dictatorships, such as Mobutu Sese Seko's Zaire or Omar al-Bashir's Sudan, all political regimes value economic growth (Olson 1993). Powerful and consolidated autocracies value economic growth because resources accruing to the regime increase. Unstable autocracies can use economic growth to bolster their support. In democracies, economic growth creates resources that the government can use to form a support base for elections. For example, an agricultural society with a shallow water table and limited access to surface water could reap major benefits from improved irrigation through pumpset electrification. In such a society, the government interest may be expected to be high, whereas a society that has abundant surface water or little prospect of improving the exploitation of groundwater would not have such an interest.

To summarize, the most fundamental condition for government action can be stated as follows: *without net political benefits to the national government, eradication of energy poverty is unlikely.* The first question any analyst of government policy to combat energy poverty must ask is whether clear political benefits, direct or indirect, are present. In the absence of such benefits, action is not to be expected regardless of other conditions.

The expectation of benefits is not enough for government action, however, and we must now consider the other pieces of the puzzle. As we saw, in the worst case, an opportunistic government offers free electricity at election time to voters. Despite the immediate relief, this practice darkens the long-term prospects of rural energy development by undermining the finances of electric utilities. Such policies may have weak, or even negative, effects on rural electrification over time if the short-run gains are set off by the long-run difficulties created by financial hardship in the electricity sector. Therefore, it is necessary to consider additional conditions that encourage interested governments to focus their efforts on effective policies and programs, instead of ineffective policies that may not achieve much more than offer patronage to political constituencies. Any given policy with the potential to eradicate poverty can be placed on a continuum from very ineffective to very effective, and we theorize about the ability of governments to enact and implement policies with high effectiveness.

We can begin by evaluating the cost of effective action by raising the question of institutional capacity. Removing energy poverty requires not only the formulation but also the implementation of new policy. However, the history of development shows that policy implementation is a major challenge in countries that do not have a long history of bureaucratic development and administrative learning (Killick 1989; Cheibub 1998). Policies that appear impressive on paper may be virtually nonexistent in practice.

Indeed, institutional capacity also plays an important role in determining the government's political strategy. Where institutional capacity is limited, the government cannot expect returns to public policies and programs that would mitigate energy poverty. Such policies would require costly investment by the government, but the lack of institutional capacity would constrain the available benefits. Therefore, in addition to undermining the implementation of government policy, the lack of institutional capacity discourages governments from policy formulation in the first place. Because a government without institutional capacity cannot expect benefits from public policy, the national leadership focuses on less productive activities, such as the distribution of patronage or repression of the political opposition.

In the case of energy poverty, the capacity required for sound public policy is found in the state apparatus in the energy sector. Relevant institutions include electric utilities, the energy and power ministry, the ministry for oil and gas, the ministry for rural development, regulatory agencies, energy cooperatives, and the rural electrification administration. There is considerable variation across developing countries in the institutional arrangements that cover rural energy affairs (Barnes 2007), and the debate on their relative merits remains unsettled. Nonetheless, no institutional arrangement for rural energy can be successful without a properly trained and motivated staff, enough budgetary resources, clear rules, and effective oversight mechanisms. Many cases in rural electrification, for example, feature institutional capacity as the key factor explaining effective implementation (Jadresic 2000; Mawhood and Gross 2014). In Nepal, Yadoo and Cruickshank (2010) attribute the achievements of the 2003 Community Electricity Distribution Bylaw, which allows rural cooperatives to make wholesale purchases of electricity and retail it to the local population, in large part to extensive government support and a strong political commitment on behalf of the national government in Kathmandu.

Modern cooking requires strong state capacity too, as new cookstoves and fuels must be disseminated. Public policies must contain credible plans to provide affordable fuel such as LPG to end users. Successful programs require long-term planning, which means that funding must be available over their entire time horizon. Joon, Chandra, and Bhattacharya (2009) examine the use of LPG in the Indian state of Haryana; they find that while many households possess an LPG cookstove, very few people use it. The main reason, they argue, is that the fuel itself ended up being too costly. This betrays the lack of long-run planning by policymakers.

The difficulties of India's Village Energy Security Programme illustrate the need for capacity building. Launched in 2004, the Village Energy Security Programme sought to increase the use of improved biomass and biogas cookstoves. Unfortunately, it met an abrupt end in 2012 when half of the commissioned projects turned out to be nonoperational. Looking at the determinants of the success or failure of each project, Palit et al. (2013) insist on the key role of institutional capacity. For instance, a lack of adequate support to train maintenance staff proved to be critical in the failure of some of these projects. Misunderstandings and lack of governmental oversight of the implementation of these projects similarly doomed these projects.

This discussion can be summarized thus: *if the government expects political benefits from eradicating energy poverty, its likelihood of success increases with institutional capacity.* In other words, conditional on a strong government interest in removing energy poverty, increased institutional capacity contributes to the likelihood of success. It does so in two ways. First, for any given level of government effort, institutional capacity improves the prospects of success. Second, institutional capacity encourages the government to expend more effort because the returns are higher.

In conducting this analysis, we recognize that institutional capacity itself depends on government interest—but only in the long run. Institutional capacity is endogenous to government interest in that governments interested in ending energy poverty are more likely than their lackluster counterparts to build capacity. This process, however, produces results only slowly, as institutional reforms move forward in fits and starts (Thomas and Grindle 1990; Levy 2014; Andrews, Pritchett, and Woolcock 2017). While we must consider the origins of institutional capacity in the long-run evolution of energy access policy, at any given time it is reasonable to hold institutional capacity constant as we investigate the government's policy choices.

Besides institutional capacity, the other factor that we expect to be central to success in the eradication of energy poverty in the presence of sufficient government interest is the strength of local accountability. According to Grant and Keohane (2005, 29), accountability

implies that some actors have the right to hold other actors to a set of standards, to judge whether they have fulfilled their responsibilities in light of these standards, and to impose sanctions if they determine that these responsibilities have not been met. Accountability presupposes a relationship between power-wielders and those holding them accountable where there is a general recognition of the legitimacy of (1) the operative standards for accountability and (2) the authority of the parties to the relationship (one to exercise particular powers and the other to hold them to account). The concept of accountability implies that the actors being held accountable have obligations to act in ways that are consistent with accepted standards of behavior and that they will be sanctioned for failures to do so.

In the case of local accountability and energy poverty, the population of the energy-poor must be able to voice their concerns and hold local government authorities accountable to their performance. As Paul (1992, 1047) writes: "Accountability in developing countries must be improved significantly to enhance the efficiency and effectiveness of their public services."

Although a national program to eradicate energy poverty requires government action, the ultimate beneficiaries of this action are local communities, and this is important for three reasons. First, local accountability reinforces the government's incentives to enact policies and implement them properly. When local accountability is strong, any policy failures will result in local complaints and protests. Second, local accountability supports and encourages the efforts of local policymakers, such as mayors of municipalities, to participate in the effort to eradicate energy poverty. To the extent that the national government can count on local support, the prospects of success are higher because local resource mobilization can supplement and complement national efforts. Finally, local accountability can contribute to the long-run benefits of eradicating energy poverty with oversight mechanisms that avoid the deterioration of service quality over time. Without local accountability, for example, the national government may implement a rural electrification program to electrify villages, but then fail to distribute electricity as power infrastructures deteriorate without maintenance.

Local accountability is not a direct substitute for institutional capacity, but it helps even if institutional capacity at the national level is low. Consider, for example, the case of a country with a very weak energy ministry and dysfunctional electric utilities that is trying to promote rural electrification. If local accountability is strong across the country, then the challenge of last-mile implementation can be safely delegated to municipal authorities. Where such authorities are capable and motivated, the weakness of the energy ministry and electric utilities may not present much of a problem. The government can invest available resources in the problem of transmission infrastructure while leaving the thorny challenge of distribution at the municipal level to local authorities. Without strong local accountability, this strategy is not possible because the government's failure to implement policy will not result in a political backlash in the municipalities. Another reason the local population's ability to enforce government policies from the bottom up is critical is information. Since the beneficiaries of energy access programs are the rural energy-poor, it would be very difficult for centralized ministries and agencies to verify the quality of implementation at the local level without some local oversight and feedback. In areas characterized by strong local accountability mechanisms, information flows from the grassroots to the national level, allowing an interested government to improve policy implementation in response to perceived problems. Additionally, local accountability allows the community to channel information about preferences, needs, and local idiosyncrasies to policymakers.

In addition to improved monitoring, local accountability is important because it ensures that the needs and preferences of the local community are communicated to policymakers and can thus inform the formulation and implementation of policy.[8]

Because the causal chain from local accountability to high-quality public service delivery is long, we would expect high levels of local accountability to produce positive outcomes only in favorable circumstances. In his comprehensive review of accountability initiatives, Fox (2015) distinguishes between tactical and strategic approaches to accountability. While tactical approaches focus only on mobilizing citizens, strategic approaches consider the role of accountability in the broader sociopolitical context. Thus, tactical approaches ignore the possibility that without changes in the political system more broadly, citizen mobilization can be ineffective because of other obstacles to improved public service delivery. In our context, local accountability mechanisms may improve information provision or mobilize citizens to demand better services, but these changes may still not reduce energy poverty unless government interest is sufficiently high. The distinction between tactical and strategic approaches to accountability thus highlights the importance of government interest as a condition for the successful use of local accountability mechanisms.

Let us illustrate the critical role of accountability with the case of Ghana. In 1990, the Ghanaian government announced the creation of the National Electrification Scheme (NES), a program designed to boost rural electrification. Its initial design was very much a top-down operation with little input from and oversight by customers. The NES focused on the most lucrative markets without consulting local stakeholders. The program was slow to make an indent in the county's low electrification rates. In an attempt to improve the results of the NES, the government augmented it with the Self Help Electrification Programme. Now, the NES would respond to demands by local households to obtain electricity. Local businesses were tasked with the construction of last-mile poles; they were also overseen by local organizers. Local accountability therefore served as a mechanism of both information transmission and to ensure the good use of the available resources. It also encouraged citizen mobilization.

In emphasizing local accountability, we must also be careful not to conflate this concept with the idea of decentralization (Bardhan 2002; Treisman 2007). Constantly debated in the development economics literature, decentralization refers to the devolution of authority and implementation capacity to the local level. As Tsai (2007) notes in her study of local public good provision in China, an authoritarian country, local accountability is ultimately about the ability of the local population to hold municipal and other local authorities responsible for their activities. Decentralization may result in powerful yet unaccountable local authorities. Such authorities have little incentive to serve the energy-poor population, and their localized power may also immunize them from top-down pressure by the government. In this regard, strong local accountability is a concept that reflects de facto accountability relationships between the population and government authorities at all levels rather than the formal division of powers between the national and local levels.

To see the difference between decentralization and accountability, consider how a recent World Bank review of the literature on the local governance of development projects highlights the importance of downward accountability in the design of development projects (Mansuri and Rao 2013, 7): "The literature is rife with cases in which decentralization is used to tighten central control and increase incentives for upward accountability rather than to increase local discretion." This observation shows that the creation of governance mechanisms at the local level may, at worst, put more pressure on local officials and populations to meet the demands of influential decision makers at higher levels of the government.

Another caveat concerns the risk of perverse forms of local accountability. In their research on decentralization and governance, with particular application to villages in rural India, Bardhan and Mookherjee (2006a, 2006b) show that decentralization carries the risk of "capture by elites." Although decentralization of decision making makes more efficient use of local information and allows the local community to use its voice to improve service delivery, this reform also enables local elites to capture some share of the benefits of public services. As Mookherjee (2015, 234) writes in a recent review of political decentralization, "Elite capture would likely depend on the extent of political competition; patterns of political awareness, participation, and literacy; and local poverty and inequality." This research attributes the advantages of the local elites to their wealth, political awareness, and connections. Because efforts to reduce energy poverty tend to focus on marginalized communities, such issues are of primary importance. In the case of rural electrification, for example, local elite capture could result in a situation wherein only a small number of wealthy, privileged households are connected to the national electricity grid. These same elites could then divert the remaining resources from household connections to their own purposes, preventing the poor from gaining access to the national electricity grid.

Even some arguments highlighting the limited effectiveness of local accountability suggest that energy poverty could be a relatively easy case for seeing positive effects. In her analysis of civil society mobilization and local governance in Mexican municipalities, Grindle (2007, 140–141) finds that citizen participation has had disappointing effects on "good governance" in general, and yet civil society groups have proven very effective in getting "government to deliver on promises to provide public works." If one accepts this generally pessimistic prognosis, the fact remains that eradicating energy poverty is to a large extent about public works. If the Mexican case generalizes to other countries, Grindle's citizen participation should prove effective in energy poverty eradication.

To understand how such accountability would work in the case of mitigating energy poverty, consider the case of a program to improve LPG infrastructure in remote rural communities. If local accountability mechanisms are strong, energy-poor people have the ability to voice their grievances and demand complete and timely implementation. For example, they could complain to municipal, or perhaps district, authorities that LPG is not available in local shops at the prices announced by the government despite earlier promises. The authorities can respond by acting to address the complaints directly or forwarding them to officials in central ministries for agencies. Without a local accountability mechanism, such complaints are ineffective. If the authorities responsible for the LPG program at the local level have nothing to fear from the population's complaints, they also have no incentive to respond to those complaints.

A particularly interesting case of the impact of local accountability is one of weak institutional capacity at the national level. In this case, the national government is largely dependent on the private sector or the local community's own efforts. As a result, local accountability influences energy poverty eradication more directly than by simply monitoring and informing government efforts at the higher levels. In Mao's China, for example, Beijing's support for rural electrification was limited to creating a conducive environment for community efforts. In Kenya, local communities in recent years have played an important role in enabling and supporting private business development in the solar sector as an off-grid alternative to grid extension.

Recent research emphasizes the possibility of strong local accountability under authoritarian regimes. According to Tsai (2007), who conducted a broad survey of 316 villages in rural China, the lack of formal accountability mechanisms has not prevented villagers in many areas from forming "solidary groups" that are able to collectively hold local policymakers accountable. As she explains, an effective solidary group must be both "encompassing" (open to many citizens) and "embedding" (incorporate local officials as members). Such a group can exercise "informal accountability" through shared norms and obligations, along with the expectation that officials are bound by them. According to Tsai's survey research and case studies, such informal accountability can explain much variation across Chinese villages in the provision of public goods and services.

In the case of an authoritarian regime with strong local accountability mechanisms, policy formulation and implementation may be quite similar to that under a democratic regime with similar mechanisms (Tsai 2007). Strong local accountability means that the government's local support base in the countryside depends on the provision of public goods and services. In the case of energy poverty, for example, local government officials would face constant pressure to improve energy access. Local leaders would then channel these demands to the higher levels of government, requesting support for local programs to improve energy access. Although democratic governments are more likely to have stronger local accountability mechanisms than their authoritarian counterparts, we shall see examples in the case studies of strong local accountability under stable autocratic rule, such as in China and Vietnam.

The determinants of local accountability are many. Political scientists and economists have emphasized factors ranging from ethnic homogeneity (Banerjee and Somanathan 2007) to social capital (Woolcock 1998) and group pressure (Tsai 2007), while adherents to the resource mobilization theory in sociology consider the presence of capable social organizations particularly important (McCarthy and Zald 1977). Since our interest is in considering the implications of local accountability as opposed to explaining variation in it, we accept these insights as valid and do not try to determine the factors driving variation in local accountability across national contexts. However, from a practical perspective, it is important to recognize that local accountability mechanisms are partially endogenous. A government committed to eradicating energy poverty can cultivate local accountability by designing robust mechanisms for enabling local participation and oversight of policies to reduce energy poverty. In the case of rural electrification, for example, we shall encounter many rural cooperatives in our case studies.

Why would governments with a strong interest in solving energy poverty ever adopt a program that does not have strong accountability mechanisms in place? If a government values local accountability, it can do its best to solve energy access problems with proper mechanisms in place, but there are several possible barriers to achieving this goal. First, the government may be unaware of the importance of local accountability. Second, it may be unwilling to transfer enforcement powers to its citizens. It may want to see energy poverty reduced, but only without the risks associated with more citizen oversight of the state apparatus. A related reason is that the government may be unable to implement accountability because local officials and bureaucrats are sometimes adept at avoiding responsibility for their actions.

More generally, local accountability has its origins in government policy. On the one hand, the building blocks of local accountability, such as institutions that empower the local population, change only slowly. On the other hand, the use of such local accountability mechanisms in the implementation of energy access policy depends on the design of the energy access policy. This combination of slow-moving and flexible drivers of local accountability means that the degree of local accountability in any given energy access policy is partially driven by the government's interest in ending energy poverty. The other major determinant is the extent to which the government can tap into preexisting local accountability systems.

While such endogenous accountability presents a difficult challenge for research design, it is important to note that the successful cultivation of local accountability inevitably requires a solid foundation. If a rural community is profoundly suspicious of the government or there are serious conflicts among the villagers, it is hard to imagine that such a community would be able to exploit local accountability mechanisms created by the government. In this sense, the successful encouragement of local accountability requires a preexisting basis.

To summarize, we claim the following: *if the government expects political benefits from eradicating energy poverty, its likelihood of success increases with the strength of local accountability mechanisms.* When government interest is present, local accountability facilitates implementation and serves as a partial substitute for national resources. Under high levels of local accountability, government policies to eradicate energy poverty have better prospects of success.

Before we move to the role of new technology, a comment on interactions between institutional capacity and local accountability is in order. As we have seen, each factor can promote success in rural electrification independently from the other. A government with high institutional capacity at the national level can implement energy access programs in a top-down fashion, whereas a government with strong local accountability mechanisms can act through municipal involvement. However, the two factors are not substitutes for each other. A government with strong local accountability, for example, performs better if it also has high institutional capacity than if capacity remains low. Therefore, while conditioning the relevance of institutional capacity and local accountability on government interest, we do not expect strong substitutability relationships between these two factors.

Finally, it is important to emphasize that the effect of local accountability —or institutional capacity, for that matter—depends on the presence of government interest. Consider, for example, a country with very high levels of local accountability. If this country has a government with a strong economic interest in rural electrification for irrigation purposes but no such interest in the provision of clean cooking fuels, we would expect to see success only in the electricity sector, whereas progress in the cooking sector would remain slow.

Role of New Technology

By now, it should be clear that fighting energy poverty is a daunting task. Over time, however, one important factor is providing governments with new opportunities for less costly and more effective action: the development of new technology for energy access. While some basic features of the energy infrastructure, such as the design of large power plants running on fossil fuels, have not changed much over time, others have undergone revolutionary changes. Such technological development unleashes new forces that facilitate the fight against energy poverty.

We argue that new technology facilitates measures to combat energy poverty across the board successfully. When government interest is low, new technology allows private markets to tackle energy poverty for commercial reasons or with international support, except in the rare case of a government that is explicitly hostile to private investment. When government interest is high, new technology creates opportunities for the government to implement less costly and more effective policies to combat energy poverty.

We first note that the development of new technologies is closely related to private markets. While the massive scale of capital investments and the limited scope for profits have historically resulted in state leadership in rural electrification and the provision of cooking fuels, recent developments in new technology have lowered barriers to entry in the field of energy access. While it would be hard to imagine a mostly commercial business model that invests billions and billions of dollars in rural energy infrastructure to serve the needs of the energy-poor, who have limited disposable income, such businesses now exist on a smaller scale. In the past, private participation in rural energy infrastructure has been mostly limited to construction contracts, with the government as the purchasing side (Hausman, Hertner, and Wilkins 2011). Today the situation is very different.

The window of opportunity opened by new technology has brought a number of entrepreneurs operating "at the bottom of the pyramid" (Prahalad, 2006) to the field of energy access. In these markets, the big challenge for entrepreneurs is the general poverty of their potential customers. For example, a study of rural energy markets and the challenges they pose to business in India lists low and volatile income, limited savings, low literacy, and diverse customer preferences as important factors (Shukla and Bairiganjan 2011, 3–4). It is only due to the dramatic decrease in the cost of new energy technology that such markets are now growing and creating opportunities for reducing energy poverty.

Even if governments, nongovernmental organizations, and other actors still must continue to subsidize technologies such as solar power, the reduced average cost of these technologies means that the unit cost of disseminating a new technology decreases. While it is true, for example, that off-grid solar technologies continue to be dependent on subsidies and government policies in many contexts (Palit 2013; Urpelainen 2014), the reduced cost of technology has allowed governments to promote the use of solar power in much larger numbers than before. Moreover, the reduced cost has allowed governments to catalyze funding from private investors.

Consider a few examples from the booming off-grid solar markets. In the case of rural electrification, there is now considerable interest in the potential for off-grid electricity provision. Private operators use solar power to generate electricity and sell it to villagers through prepaid systems, often exploiting mobile technology to reduce the transaction costs. Others sell solar home systems funded with soft loans, which offer favorable financing conditions to borrowers. Manufacturers develop new biomass cookstove models that are more efficient, convenient, and attractive to the customers. In some cases, these new technologies have been deployed on a large scale. Bangladesh is a case in point. Average income is extremely low, and about one-third of the population lives in poverty. A large proportion of the population also lives in rural areas, which are, as we know, hard to reach and complicate electrification efforts. Despite these obstacles, solar home systems have proved to be an impressive success. By most accounts, more than 480,000 solar home systems were used in Bangladesh by the early 2010s (Samad et al. 2013, 3).[9] And while this number is dwarfed by the total population in dire need of access to electricity, it underscores the remarkable progress made by new entrants in this market.

The rapidly emerging field of commercial off-grid electrification also provides several illustrations of how such businesses operate. One of the well-known examples is undoubtedly Grameen Shakti, a Bangladeshi firm founded in 1996 by Muhammad Yunus, who would later win the Nobel Peace Prize (Samad et al. 2013, 5). Grameen Shakti is at the forefront of the push for solar home systems in rural areas. Its business strategy is simple (Amin and Langendoen 2012). Grameen Shakti sells solar home systems that are technologically simple and therefore cheap. The company then relies on its deep ties to local communities to find buyers. If needed, consumers can take loans from Grameen Shakti; this is important to attract cash-strapped consumers. Furthermore, Grameen Shakti provides cheap maintenance. Poor maintenance services are often a key reason for prospective buyers to favor alternative options. And according to its internal numbers, Grameen Shakti is successful, passing the 1 million solar home systems mark in 2012 and standing at over 1.5 million according to the latest data.[10]

Similar examples are now found in the field of clean cooking solutions, though overall progress has been slower. In India, for example, the company SustainTech designs and markets clean cookstoves. In a September 2014 interview with *Times of India*, the company's chief executive officer, Svati Bhogle, estimated that the company had sold 3,000 stoves in its first three years of operation and was aiming for annual sales of 5,000 by 2016.[11] Major multinational corporations such as Philips have begun to explore the cookstove market in sub-Saharan Africa.[12]

What is the significance of these developments for the mitigation of energy poverty? Most fundamental, they reduce the cost of improving energy access for the government. New technology expands the range of options available for offering modern energy access to the rural energy-poor. If the government so wishes, it can still continue to rely on traditional approaches, such as grid extension for electricity generation from, say, coal power plants. But some governments also explore the possibility of relying on new technologies. While these new technologies may not completely replace traditional approaches, they are a useful addition to the government's tool kit. For example, the government could emphasize grid extension in communities that are not located far from the existing grid while also supporting the construction of solar power plants in more remote communities. In Bangladesh, Grameen Shakti specifically targets communities that are unlikely to get access to the grid within the next five to ten years (Kaneko, Komatsu, and Ghosh 2012, 211). With support from the government and donor agencies, notably the World Bank, Grameen Shakti is able to fill rural electrification gaps in areas beyond the reach of the government's grid extension efforts.[13]

The importance of new technologies is clear in the predictions of the IEA's 2017 *Energy Access Outlook* (IEA 2017). In the IEA's New Policies scenario, global electrification rates increase from 86 to 92 percent as Africa remains the only region with a large nonelectrified population (36 percent of total population). Unlike in the past, however, the IEA predicts mini-grid and off-grid solutions to offer access to two-thirds of the rural population that gains access to electricity. In the more aggressive Energy for All scenario, this number is even higher as more aggressive policies contribute to investment in distributed power generation in the rural areas. If the IEA's predictions prove correct, the future of energy access depends on a different set of technologies—and therefore a different set of policy challenges.

In all likelihood, however, the growing commercial sector that exploits new technology cannot be a substitute for the government. The creation of private markets is an activity that depends on the regulatory environment. A government that is unable to guarantee basic rule of law and security of property rights cannot expect robust private markets for energy access. Similarly, a government that adopts an overly aggressive regulatory approach, requiring cumbersome licensing procedures, can basically grind private business to a halt. At the very least, the government must be able to guarantee a basic environment for businesses. If it wants to rely more heavily on the private sector, it can also form public-private partnerships.

To illustrate, consider the problem of regulating an off-grid electricity generation system in a remote rural community in India. When we discussed these challenges with local entrepreneurs in New Delhi in spring 2014, they told us they had several concerns about government action. On the one hand, they were concerned about proposals to regulate off-grid tariffs. If the government were to impose a cap on electricity prices from off-grid systems, the entrepreneurs said, it would be very hard to break even in communities that need better lighting the most. On the other hand, the entrepreneurs also acknowledged that they were dependent on the government for the protection of property rights. This was the time of the Indian general elections that brought the Bharatiya Janata Party (BJP) candidate Narendra Modi into the prime minister's office in a landslide, and some Congress politicians had forced off-grid entrepreneurs to stop collecting payments during the election period in what proved to be a futile attempt to build popularity in the villages.

Since the potentially powerful combination of new technology and private entrepreneurs does not remove the need for government action, government interest in removing energy access remains a key predictor of success regardless of the level of technology development. New technologies relax resource constraints and create new opportunities, but they do not remove the fundamental need for investing in infrastructure. In this sense, energy is very different from telecommunications, where the low capital requirements have allowed private entrepreneurs to create a robust infrastructure for mobile phones with virtually no government support.

Institutional capacity also remains critical, though the challenge of using new technologies and engaging the private sector is different from the traditional challenge of capital investments in massive public works. In a government-led effort, institutional capacity is required because the government itself must construct infrastructure or, at the very least, cooperate with large corporations that are capable of implementing large construction projects. In a strategy that emphasizes private business development, the challenge is now appropriate regulation. Specifically, the government needs to create what many entrepreneurs call an "enabling environment" for their activities. The local rules and regulations must enable profitable business, while ensuring that entrepreneurs do not engage in predatory pricing and that their profit-motivated activities ultimately benefit the energy-poor.

For example, a government that engages in heavy-handed regulation may stifle the development of off-grid rural electrification (Tenenbaum et al. 2014). If the government regulates the electricity prices for off-grid operators or imposes cumbersome licensing requirements, small business may be unable to remain profitable. Conversely, the lack of any kind of regulation could also create problems. For example, the provision of subsidies to allow the energy-poor to gain access to solar power inevitably requires some rules for acceptable technical specifications. If there are no such rules or they are not enforced appropriately, the abuse of the subsidy regime is a threat to the success of the policy.

The importance of local accountability also remains strong. As private entrepreneurs enter rural energy markets at the bottom of the pyramid, they have to engage the local society. If local accountability mechanisms are strong, for example, the rural energy-poor can demand that their municipal authorities allow energy access entrepreneurs to operate and do not stifle business through corrupt or nepotistic practices. Conversely, local accountability means that entrepreneurs have an incentive to conduct fair business and improve their social impact, because the energy-poor can complain to authorities in case entrepreneurs initiate predatory pricing, try to prevent competition, or otherwise abuse the local population. Tenenbaum et al. (2014, 78) use Cambodia as an example of the benefits of local accountability. In this country, a private power generator signed a fifteen-year electricity supply agreement with the local government of the Smau Khney village. The contract specifies the terms of electricity provision, such as prices and hours of supply, and offers a modest annual budget for the village electricity committee's activities.

To summarize, while new technology is not a panacea for the problem of energy poverty, it can prove helpful in a wide variety of national circumstances. New technology is not a substitute for government action, institutional capacity, and local accountability. If the conditions are not conducive to decisive government action, the private market provides relief in the shadow of the government, and such relief may be significant since the government has yet to achieve much. But if the government has the willingness and ability to act, then new technology creates opportunities for faster progress, relative to an already good baseline. *New household energy technologies could increase the likelihood of success in the eradication of energy poverty for all developing countries. This is true for any level of government interest, institutional capacity, and the strength of local accountability mechanisms.*

Contrasting Rural Electrification and Modern Cooking Fuels

While both rural electrification and the provision of modern cooking fuels are vexing issues, an important implication of our theory is that we should generally see more government interest in rural electrification than in provision of access to modern cooking fuels. Compared to improved access to cooking fuels, rural electrification promises many more direct benefits, both political and economic, to the government. Government interest is the critical precondition for success, and rural electrification is simply more attractive to developing country governments than investment in clean cooking.

Despite the high capital intensity of rural electrification through traditional grid extension, this strategy is attractive for governments. First, it often promises direct productive benefits. Studies of rural electrification show that it contributes not only to agricultural productivity but also to nonagricultural economic activity (Cabraal, Barnes, and Agarwal 2005; Cook et al. 2005; Bernard 2010; Dinkelman, 2011). In India, access to electricity enables the use of electric pumps, which improves irrigation (Sekhri 2011). In turn, farmers can then grow crops that are more valuable and therefore increase their returns. Similarly, rural electrification empowers small businesses, such as grocery stores, which rely on operating after sunset and on items such as refrigerators (World Bank 2008, 47). Sometimes electrification also facilitates productive activities on a larger scale, such as rural factories.[14]

In contrast, access to clean cooking often does not have similar direct economic benefits. While clean cooking can undoubtedly improve health outcomes within households (Kandpal, Maheshwari, and Kandpal 1995; Bailis et al. 2009), these improvements add to productivity only indirectly and over time. A successful rural electrification program, in contrast, can improve agricultural harvests immediately through better irrigation, whereas clean cooking does not provide such direct benefits. Even benefits such as reduced deforestation are unclear, as the government may not ascribe a premium to forest protection. Additionally, the use of biomass for cooking need not cause deforestation if people rely on branches and deadwood instead of cutting down healthy trees (Arnold, Köhlin, and Persson 2006).

Moreover, the evidence shows that the demand for clean cooking among households is much less pronounced than the demand for electricity access. While LPG is a convenient alternative to biomass, the cost of first purchasing an LPG stove and then incurring fuel expenditures is high, and few rural households in poor countries can afford to do so. In India, for example, the penetration of LPG remained at only 24 percent in 2010. Worse, of these households, only about one-third relied on LPG exclusively, meaning that increased LPG use would not even solve indoor air pollution. In Bangladesh, Mobarak et al. (2012) show that people are willing to pay virtually nothing for efficient cookstoves. Experiences with clean cooking programs show that the design of efficient cookstoves is difficult, and while people value basic electric appliances almost universally, their cooking preferences vary greatly across localities (Lambe and Atteridge 2012; Sehjpal et al. 2014).

Political-economic considerations reinforce the advantage of rural electrification. Since rural electrification allows enhancing productivity more than clean cooking does, the potential coalition in favor of rural electrification is larger than the coalition in favor of clean cooking. For example, both major agricultural producers and industrialists with existing or planned facilities outside major urban centers may expect concrete benefits from rural electrification. Government officials, in turn, may be interested in the rents available from major infrastructure projects, such as the construction of power plants, transmission lines, and distribution infrastructure (Wade

1985). These elite interests have economic and political resources not available to the rural-energy poor. As Kale (2014a) shows in her analysis of the origins of rural electrification in Indian states, state governments sowed the seeds of enhanced electricity access decades ago in their effort to capitalize on the Green Revolution through electricity supply for irrigation. And yet rural electrification projects are also highly visible, with measurable outputs, and thus allow governments to claim credit for them.[15]

In contrast, the provision of clean cooking fuels does not allow such coalition building. Because clean cooking fuels in most cases do not have direct productive uses, the political coalitions that can be built to support them do not include the domestic economic elites. As we shall see in our case studies, governments are much less interested in the public health and environmental benefits of replacing traditional cookstoves with alternatives than in the direct economic benefits of electricity supply. Of course, infrastructure contractors have an interest in both rural electrification and clean cooking programs. Because the interest is shared, however, it does not drive a wedge between the political feasibility of the two policies. Any political differences between the two must be attributed to other factors, such as expected benefits to domestic economic elites.

A historical condition that further amplifies the difference between rural electrification and clean cooking is that major international donors have put more emphasis on rural electrification, presumably because of the promised productivity improvements. The World Bank, for instance, regularly spends large amounts on electricity-related projects. In 2017, it spent a bit less than US$5 billion on electricity and transmission projects (separate spending on rural projects is unfortunately not available).[16] An internal assessment by the World Bank (2008, 10) on its spending over the past three decades offers a "lower estimate [of] $798.3 million from 1980 to 2006. An upper estimate ... comes to $5.97 billion." Similar numbers are not available for cooking projects. However, an AidData project sheds some light on the discrepancy between electrification and cooking. AidData collects foreign aid data from a number of countries and categorizes projects based on their goals. Summing up spending across the world on projects that have a rural electrification component, we find that donors spent an estimated US$36 billion between 1949 and 2013, or about US$562 million per year.[17] Projects for cooking amounted to only US$1.2 billion over the shorter 1966–2013 period, which amounts to about US$25 million per year. Even if the data suffer from measurement errors, the difference in energy access efforts between electrification and cooking remains substantial.

Drawing these arguments together, government interest can be expected to be larger in the case of rural electrification than in the case of modern cooking fuels. This does not mean that governments are never interested in modern cooking fuels, but we do expect such interest to be present more often in the case of rural electrification. As a result, variations in institutional capacity and local accountability should also be more strongly associated in the case of rural electrification. When government interest is low, as is often true for access to modern cooking fuels, variation in institutional capacity and local accountability may not explain much because the lack of government interest raises insurmountable obstacles to success. In the case of high government interest, however, variation in institutional capacity and local accountability is critical to explaining the difference between successful and failing policies, programs, and projects.

To be sure, we do not claim that modern cooking fuels are insignificant. To the contrary, we wrote a book about them because we believe their provision is fundamental to public health, quality of life, and the environment in developing countries. However, the harsh reality is that the provision of modern cooking fuels is not very attractive for national governments. As a result, there will be a gap between rural electrification and the provision of modern cooking fuels. It is only in rare circumstances, such as the case of Senegal, that we see more progress in the provision of cooking fuel than in the provision of electricity access. We do observe, however, that some of the countries that solved the electricity problem later also solved their cooking problem. China and South Africa, for instance, achieved this over the past few decades.

Case studies are necessary to understand the actual processes that contribute to improved energy access. The next three chapters present in-depth analyses of specific countries. In chapter 4, we examine India, a country that has experienced radical changes since its independence in 1947 and warrants its own chapter for a number of reasons. First, its federal system offers a particularly interesting test for our hypotheses, because it puts us in a position to explain both successes and failures within a single country. Second, solving energy poverty in India is one of the most pressing policy challenges, given the country's huge and rapidly growing population. Finally, our earlier work in India offers special insight into the problems faced by this country. In chapter 5, we present countries that have recently been successful in solving their most pressing energy poverty issues. Finally, in chapter 6, we examine countries that still face tremendous obstacles in providing a minimum amount of energy to their most vulnerable populations.

The strengths of case studies are manifold. First, they reduce problems of measurement. Institutions that shape local accountability can take various forms. By looking at single countries, we can drop generic proxies that need to hold equally across the world and instead adjust our measurement according to the national context. Accountability in a country such as India may look very different than it does in, say, China. The intricacies of the political process are therefore easier to capture if we do not have to apply the same grid, ex ante, to all countries. Similarly, we have highlighted the difficulties in measuring government interest. The presence of democratic institutions is a good predictor of interest, but we also recognize that it is a blunt measure. A case study that mixes qualitative and quantitative evidence enables us to triangulate by drawing on a larger body of evidence to assess the presence or absence of government interest.

Second, case studies are less restrictive in terms of exact model specification. For instance, it is not always clear how quickly we expect the effect of changes in our key variables to materialize. Suppose a country begins a democratization transition; our theory predicts that electrification rates ought to increase, but this may take time. Some countries can rapidly mobilize the means required to implement their policies; others may take longer.

Now take the example of accountability. Imagine that government officials reform the way local politicians are selected or removed from power. The time for this institutional reform to change the incentives of policymakers may vary considerably across countries. In some countries, a particularly public removal of a well-known official may deter future policymakers from poorly executing policies. Elsewhere, the reform may not worry local officials for a long time. More generally, by looking at individual countries, we reduce concerns about the appropriate parametric structure of the relation between politics and energy poverty. We do not have to know ex ante how our variables affect energy poverty.

When looking at the evidence provided in country-specific studies, we need to ensure that we can trace the causal process closely. Process tracing requires the following steps: "One carefully maps the process, exploring the extent to which it coincides with prior, theoretically derived expectations about the workings of the mechanism. The data for process tracing are overwhelmingly qualitative in nature, and includes historical memoirs, interviews, press accounts and documents" (Checkel 2005, 6). In this respect, a critical consideration is the selection of cases (Achen and Snidal 1989; Gerring 2004; Levy 2008; Seawright and Gerring 2008). As noted, we must avoid choosing countries just because they fit our argument. Instead, we need a systematic sampling strategy.

In sampling, we combine two approaches from the qualitative literature. First, we select individual countries based on the "most different" approach (Seawright and Gerring 2008). By this, we mean that we want to vary as much as possible the cases based on our control variables. Typically, this ensures that the causal relation between our political variables and energy poverty is not confounded by other variables. For instance, we wish to show that our political variables operate in both small and large countries. This would suggest that our findings are not driven by these third variables. In practice, we sampled countries from all continents except Europe and North America to reduce concerns about the role of natural resources and geographical latitude. We also let country size vary, with cases ranging from giants like Nigeria to much smaller ones such as Ghana. Finally, we picked both wealthy countries and poor ones. Brazil, for example, is much wealthier than Vietnam. Overall, we therefore drew on evidence from a wide variety of settings among both the success cases and the failures.

To make causal inferences about the effect of our political variables, we rely on a second central principle of qualitative methods: the "most similar" approach (Seawright and Gerring 2008). To identify the effect of political factors such as democracy or capacity, we turn away from cross-national comparative studies. Instead, we rely on within-country changes. In other words, we compare countries (and in some cases subnational units) both before and after they experienced a change in one of the three facets of our theory: government interest, institutional capacity, and local accountability. Assuming that other variables, such as population density, change only slowly over the same period, this approach enables us to uncover the effect of the variable in question. Relying on these guidelines, we selected the following countries. In chapter 5, we discuss the success cases: China and Vietnam (Asia), South Africa and Ghana (Africa), and Brazil and Chile (Latin America). In chapter 6, we present countries that faced challenges and discuss cases in greater depth: Bangladesh and Indonesia on the Asian front and Kenya and Nigeria in sub-Saharan Africa. We also consider the deviant case of Senegal because it presents the rather unique scenario of rapid and early progress in clean cooking fuels without corresponding advances in rural electrification.

4 Persistent Energy Poverty in India

Rural energy access has been a vexing problem for India ever since the country's independence in 1947. Despite numerous initiatives over the past seventy years, more than 250 million people in India still remain without basic electricity access and 870 million continue to use biomass as their primary cooking fuel. The provision of electricity and clean cooking fuel access to rural households has faced numerous problems, though rural electrification in India has seen some progress over recent decades. When the country gained independence, less than 1 percent of villages were electrified, and total power output was a little more than 1 gigawatt (Samanta and Sundaram 1983; Government of India 2017).[1] By 2010, around 79 percent of the total population consumed electricity (93 percent in urban areas and 73 percent in rural areas), and by 2013, the installed capacity had increased to 223 gigawatts (NSS 2012a; Government of India 2017).[2]

This increase masks the fact that nearly 225 million people living in rural areas still do not have access to electricity.[3] The growth in overall rural electrification rates has also been uneven. Some states, like Gujarat, Tamil Nadu, and Punjab, have achieved over 90 percent rural electrification, but others, like Bihar, Uttar Pradesh, and Assam, suffer from rural electrification rates as low as 37 percent (Government of India 2011c).[4] Moreover, much of the increase in rural electrification was achieved in the past fifteen years: at the turn of the century, less than 50 percent of rural areas had access to electricity (NSS 2012a). Despite numerous studies confirming the benefits of providing universal electricity access and the harmful effects of using kerosene as a lighting fuel (Khandker, Barnes, and Samad 2012; World Bank, 2008; Lam, Smith, Gauthier, and Bates 2012), India's initiatives to provide electricity access to rural areas can at best be deemed as a partial success.

The provision of clean cooking fuel in India has clearly been less successful than that of electricity access. In 2010, around 72 percent of the population still used biomass fuels, this number rising to almost 90 percent in rural areas (NSS 2012a). Although the harmful effects of biomass are well known (Barnes et al. 2009; Pandey et al. 1989; Smith 2000; Schei et al. 2004; Epstein et al. 2013), there has been only a small decrease in its use since these assertions were accepted. Compared to the late 1980s, reductions in the use of biomass were minimal, with 13 percentage points in the overall use of biomass and a minimal 3 percentage point drop in rural areas. While there has been a modest increase in the use of cleaner fuels like LPG, these have mainly benefited urban households (Rao 2012). Many of the government's efforts have focused on providing people with an intermediate solution in the form of improved cookstoves; these devices still use biomass but attempt to curb the level of indoor air pollution. However, the uptake of these stoves has been low and has led to fuel stacking, with households using both the traditional and modern alternatives (Cheng and Urpelainen 2014). National efforts to provide people in rural areas with access to clean cooking fuels have been halfhearted at best. Though there have been some initiatives in recent years to provide access to clean energy fuels, it is too early to judge whether they will achieve their intended goals. Overall, the Indian government's efforts to provide rural areas with access to clean energy fuels have been an unequivocal failure.

In this chapter, we explain the reasons behind the continued energy poverty in rural India, focusing on both temporal and spatial variation within the country. We show that existing explanations of energy poverty such as resources, geography, and wealth do not adequately explain the variation in access to electricity and modern cooking fuels. In addition, we show that explanations focusing specifically on energy poverty in the Indian context, such as ineffective policies, inefficient governance, decentralized options, and financial incentives, are insufficient for an in-depth understanding of the situation. We then describe how our theory outlined in chapter 3 provides a better explanation of energy poverty in India. To test this theory, we evaluate India's progress over time and across states. We show that government interest in rural electrification has increased over time, especially because India is a democracy and rural electricity supply became key to exploiting the country's groundwater resources for agriculture. Coupled with continued improvements in institutional capacity, we argue that these changes in India's political economy have enabled progress, but the lack of local accountability has limited the pace of this achievement. Leveraging India's federal structure, we consider variation across five states—Orissa, Uttar Pradesh, West Bengal, Gujarat, and Maharashtra—and show that state-level variation in government interest, institutional capacity, and local accountability can explain progress in rural electrification in these regions.

Consistent with our expectations, progress in the provision of modern cooking fuels has been much slower. We show that this lack of success can be attributed to low levels of government interest. Our treatment of cooking fuels is more cursory than rural electrification because the lack of government interest can explain India's uniformly disappointing record in this sector.

Before we embark on a detailed discussion of energy poverty in India, it is useful to understand why we focus on this country. First, India is a large country, as we have noted, and it provides us with considerable variation over both time and space to evaluate our argument. While the case studies in chapters 5 and 6 are useful to understand the relevance of our theory at the national level, this chapter allows us to examine our argument at both the national and subnational levels. Second, the country has a large rural population, which allows us to test whether our theory holds for a large proportion of people without electricity or modern cooking fuel access. Finally, India is a typical developing country, and understanding energy poverty here will allow us to draw lessons for other countries in the region and beyond.

Next, we describe India's energy poverty problem and consider the strengths and limitations of existing explanations. The following two sections offer an in-depth analysis of rural electrification in India, first at the national and then at the state level. The provision of modern cooking fuels is considered only briefly, given that the explanation turns out to be straightforward. The final section discusses the future of energy access in India.

India's Energy Poverty Problem

India's energy poverty problem has deep historical roots. To set the stage for our analysis, we first discuss the evolution of access to electricity and modern cooking fuels over time.

Electricity Access

To understand the breadth and depth of rural electrification in India requires us to start at a time when it was still under British colonial rule. During this period, electricity was supplied through private companies as dictated by the Indian Electricity Act of 1910. These companies were primarily located in cities and large industrial towns, with the British provinces of Bengal, Bombay, Madras, and Mysore accounting for a vast proportion of the total installed capacity (Rao 2010). Kale (2014b) believes that the initial colonial emphasis on these areas played a significant role in the functioning of the modern Indian electricity sector; for instance, the involvement of private companies in electricity generation and distribution in places like Bombay limited the ability of the state to enter the electricity sector until gaining independence.[5] Most important, the focus on these urban centers resulted in the neglect of rural electrification. Only about six out of one thousand villages were electrified when India became independent in 1947, and the country had a total power output of only around 1 gigawatt (Samanta and Sundaram 1983; Government of India 2017). This is approximately equivalent to the output of just one unit of a typical nuclear reactor in the United States.

The role of electricity in uniting the nation was not lost on the first generation of Indian leaders (Kale 2014a); both Mahatma Gandhi and Jawaharlal Nehru were interested in identifying ways to unite the country after the British left, and expansion of electricity access was viewed as a critical factor in achieving this objective. Similarly, B. R. Ambedkar, the chairman of the committee on public utilities and electricity at the time, espoused the need to provide cheap electricity to everyone in the country. While the reason calling for the expansion of electricity access at the turn of independence was one of unity and industrialization, India was also a democratic country with regular elections, and providing electricity access was in the interest of the national government at the time. In other words, nationalistic ideals aside, there was a clear political incentive, that is, government interest, to address the problem.

Given that a more-or-less tabula rasa environment existed in 1947, at least as far as rural electrification is concerned, India has made modest progress over the past seventy years.[6] In 2010, about 79 percent of the population had electricity access, though that number was around 73 percent in rural areas. Moreover, the country had a generation capacity of about

223 gigawatts by March 2013 (NSS 2012a; Government of India 2017). Figure 4.1 shows the growth in the number of electrified villages and per capita consumption of electricity between 1950 and 2012. The left panel shows a steady increase in the number of electrified villages over time with the rate increasing sharply after the mid-1960s.[7] Similarly, the right panel shows a steady increase in the proportion of rural households with access to electricity.

Considering where India began at independence, this may seem like a considerable improvement. However, these numbers mask the fact that in 2010, over 250 million people in the country, about 90 percent of them living in rural areas, lacked access to basic electricity. The situation was worse just a decade earlier: around 411 million people did not have electricity access, of whom 371 million lived in rural households. The increase in overall electrification rates over the past seven decades also belies another important fact: the vast number of nonelectrified households are not evenly spread across the country. Figure 4.2 shows the spatial variation in levels of rural household electrification. In 1987, northwestern states like Rajasthan and Haryana achieved more than 50 percent rural electrification, but the southern state of Andhra Pradesh was the only one with an electrification rate of more than 75 percent. In 2010, the western and southern states of Gujarat and Tamil Nadu electrified nearly 95 percent of rural households, while the northern state of Bihar was languishing at less than 35 percent in rural areas.

While there has been a steady increase in electricity consumption over time, the right panel in figure 4.1 also shows that kerosene consumption has decreased only slightly over the last two decades: in 2010, 88 percent of rural households still used kerosene. In sum, rural electrification rates around the country have improved over the past seventy years, but millions still lack basic electricity access. Moreover, some states in India have lagged behind others in providing reliable rural electricity (Aklin et al. 2016). Together, these numbers present a snapshot of the massive rural electrification challenge that still lies ahead.

Cooking Fuel Access

Access to clean cooking fuels in India has been and continues to be a problem for the government. Typically, Indian households use two types of cooking fuel: LPG as a liquid fuel and biomass as a solid fuel. The cleaner

a)

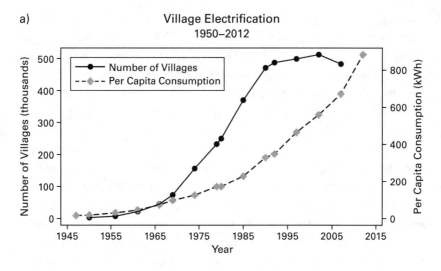

b)

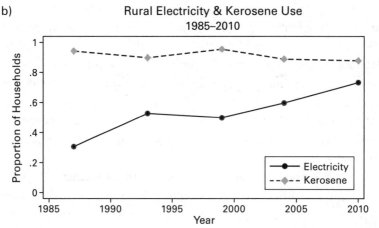

Figure 4.1
Village and rural household electrification. a) shows the increase in the number of electrified villages (left axis) and per capita electricity consumption (right axis). Per capita consumption is the gross electrical energy available divided by the mid-year population. b) shows the proportion of households having access to electricity (straight line) and using kerosene (dashed line) as their main cooking fuel. Source: Village electrification data are from Government of India (2017). Household-level electrification data come from the National Sample Survey, 1987–2010.

Rural Household Electrification

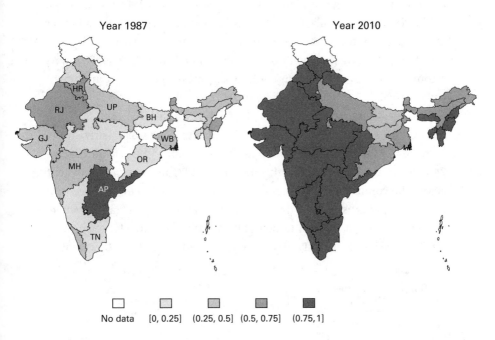

Figure 4.2

Rural household electrification by state, 1987 and 2010. Electrification is defined as nonzero electricity expenditure in the past thirty days at the time of survey. Note that this is a lax definition of electrification, as it could also include payments for power from a local diesel generator. The two-letter codes displayed on the Year 1987 map reflect the states we discuss in this chapter: Rajasthan (RJ), Haryana (HR), Andhra Pradesh (AP), Gujarat (GJ), Maharashtra (MH), Orissa (OR), Uttar Pradesh (UP), West Bengal (WB), Tamil Nadu (TN), and Bihar (BH). Source: National Sample Survey, 1987 and 2010 rounds.

fuel, LPG, is primarily used by richer and more urbanized Indian households (Balachandr 2010). In 2010, around 67 percent of urban households used LPG compared to just 24 percent among rural households. Rural areas of the country still use biomass through cookstoves called *chulhas*. These devices use biomass fuels like wood, crop residue, and dung and generate a lot of smoke and soot—so other than using an energy-inefficient fuel, these chulhas cause high levels of indoor air pollution. The harmful effects of such indoor air pollution have been well documented in India, as well as other developing countries (Smith 2000; Pandey et al. 1989; Mishra,

Retherford, and Smith 1999). Moreover, India has faced a firewood crisis because of vanishing forest cover since at least the mid-1980s (Patel 1985; Baidya 1986). As a result, rural household members have to travel farther and longer to retrieve the necessary firewood for use in their chulhas, or switch to collecting firewood from private trees (Van't Veld et al. 2006). As a consequence, even rural households with access to LPG use a mixture of cooking fuels (Cheng and Urpelainen 2014).

Figure 4.3 presents the number of rural households that used LPG and biomass from 1987 to 2010. On a positive note, there was a fairly steady increase in LPG use among rural households, from 5.5 percent in 1987 to 24 percent in 2010. However, this increase in LPG use was not consistent across the country. In 1987, almost all states had less than 20 percent of rural households using LPG; the only exception was Andhra Pradesh, where around 27 percent of rural households said they use the cleaner fuel. In 2010, Himachal Pradesh, Punjab, and Kerala had more than 50 percent of

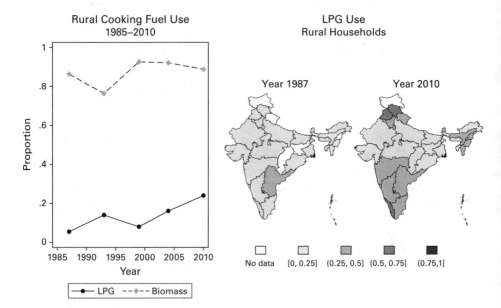

Figure 4.3
LPG and biomass use, 1987–2010. LPG and biomass use are defined as nonzero expenditure in the past thirty days at the time of survey. Because biomass can be collected without a monetary payment, the value of own production is imputed for households that report self-collection. Source: National Sample Survey, 1987–2010.

rural households using LPG. Other southern states like Tamil Nadu, Karnataka, and Maharashtra and northeastern states like Sikkim, Manipur, and Mizoram also made progress with more than 25 percent of rural households using the fuel for cooking. However, rural households in Chhattisgarh, Jharkhand, Orissa, and Bihar continued to have low numbers of households that used LPG. Importantly, the increase in LPG use between 1987 and 2010 was not accompanied by a decrease in the use of biomass. Over 86 percent of rural households in India were using biomass in 1987, and around 89 percent still continue to do so today.

Existing Explanations for India's Energy Poverty Problem

There are clear benefits to improving energy access (Dinkelman 2011; Barnes, Van der Plas, and Floor 1997; Khandker, Barnes, and Samad 2013; Barnes et al. 2009), and these have been demonstrated in the Indian context. For example, Khandker et al. (2012) find that the provision of electrification to rural households increases labor supply, household per capita income and expenditure, study time, and schooling outcomes for both genders, and it also reduces time spent collecting firewood. In a recent study, Banerjee et al. (2015) show that electrified Indian households spend less per unit of lighting than do those who use kerosene. They also show that for a fixed amount spent on light, electrified households enjoy more than one hundred times more light. There is evidence in the Indian context that people who live in households that use kerosene are associated with increased risk of low birthweight babies (Epstein et al. 2013).

Better access to clean cooking fuels also has clear benefits. The use of biomass cookstoves results in higher levels of indoor air pollution. Many studies document its negative health effects and how its use is a significant environmental and public health challenge (Barnes et al. 2009; Pandey et al. 1989; Smith 2000; Schei et al. 2004). The use of biomass cookstoves in the Indian context has been associated with tuberculosis, for example (Mishra, Retherford, and Smith 1999). Moreover, evidence suggests that biomass is associated with low birthweight and neonatal death (Epstein et al. 2013), and women and children are more likely to experience the negative effects of indoor air pollution from the use of these biomass cookstoves (Parikh, Smith, and Laxmi 1999).

If there are such clear benefits to the use of both electricity and clean cooking fuels, why has India faced steep problems in providing them? The vast majority of the energy access literature on India tends to focus on either the problems associated with rural electrification or with the provision of clean cooking fuel. While these approaches cannot directly address the overarching problem of energy access as a whole, the existing literature is useful in understanding issues that surround each specific sector. We begin by critically examining this literature and discuss the strengths and limitations of their solutions to improve energy access in India. Besides the fundamentals of geography, energy resources, and income, the literature has typically focused on inefficient policies and lack of good governance. Some scholars have also noted the potential of decentralized solutions for electricity access and cooking, but none of these solutions respond to the wide variation that currently exists across states, and indeed they may not contribute toward progress without fundamental political changes. A few exceptions notwithstanding, most of this research ignores the political-economic dimension of the problem of energy poverty. In short, individual approaches provide useful insights into specific dimensions of energy poverty, but they are inadequate on their own to explain all of the changes across space and time.

Geography, Energy Resources, and Income

As we discussed in chapter 2, geography, resource availability, and income are important conventional explanations for explaining variation in energy access worldwide. In the context of rural India, these explanations have some use but are not sufficient to explain the variation in electricity and cooking fuel access. We consider each of them in turn.

First, geography could play a role in explaining why the Indian government did not expand its rural electrification policies. Suppose a particular state had a number of hilly regions or low population density; the government might not want to invest in the necessary infrastructure for such places. This immediately suggests that places with higher population density should have higher levels of rural electrification given the ease of access to and the demand for electricity in these areas. However, states like Bihar and Uttar Pradesh have, respectively, the sixth and eighth highest population densities among Indian states and union territories, and yet have some of the lowest rural electrification levels in the country. According to the

2011 national census (Government of India 2011a) and National Sample Survey in 2010 (NSS 2012a), Bihar has a density of 1,106 people per square kilometer with rural electrification levels of just 34 percent, whereas Gujarat has a population density of just 308 people per square kilometer and yet has a rural electrification rate of 95 percent. Similarly, Uttar Pradesh is not as hilly as the coastal state of Maharashtra but has a rural electrification rate of 54 percent compared to over 90 percent in the latter area. These percentages suggest that geography and low population density do not provide an adequate explanation of rural electrification rates in the country. Geography also does not explain why rural households still continue to use biomass since wood, crop residue, and dung are all locally available and are independent of the topography of a particular location. Indeed, some of the hilliest Indian states, like Arunachal Pradesh and Karnataka, have similar biomass use rates in rural areas compared to Madhya Pradesh and Orissa (NSS 2012a).

Another possible explanation is that resource availability makes it easier for some places to have better rural electrification rates than others. According to this explanation, states that have greater energy resources would have the capacity to generate electricity more quickly and would incur lower transmission and distribution costs than other states. As a consequence, these areas should have higher rural electrification rates than places without access to such resources. In India, the source of electricity generation is primarily through coal, and if the resource availability explanation is valid, we should see evidence of major coal-producing states faring better in terms of rural electrification. However, the eastern state of Jharkhand has nearly 40 percent of the country's coal reserves and the highest coal production, and yet rural electrification rates stood at a measly 48 percent in 2010. Consider another example: Orissa's coal production is around 13 percent of the country's total output, and yet 66 percent of its rural population have electricity access. In short, the overall evidence suggests that resource availability is not a good explanation for rural electrification levels in India. Furthermore, it does not account for the continued use of biomass; if this explanation were to hold, we should see lower biomass use in places where these resources are scarce. However, Uttar Pradesh and Orissa, among the states with the lowest forest cover, have higher biomass use rates in rural areas compared to states with more forest cover like Himachal Pradesh (NSS 2012a). Moreover, Van't Veld et al. (2006) show that in places where firewood is scarce, landed households switch to agricultural waste, and landless households begin using firewood from private trees; hence changes in the availability of resources like firewood do not automatically lead to a decrease in the use of biomass fuels.

A third possible explanation is about income. India is a relatively poor country with a per capita GDP of less than US$1,500. According to the income explanation, richer states should be able to provide better electricity access to its rural households. This explanation does better than geography and resource availability since richer states generally do have better electrification records. For instance, Gujarat and Maharashtra achieved at least 70 percent rural electrification levels by 2010, while poorer states such as Uttar Pradesh and Bihar lag behind considerably (Government of India 2011c). This explanation is also insufficient because it is unable to explain rural electrification levels in states like Andhra Pradesh. In 1987, Andhra Pradesh was among the poorest Indian states, but it was the only one achieving rural electrification rates of more than 75 percent (NSS 2012a). The income explanation performs reasonably well in the context of cooking fuels, however. Given the higher price of liquid, nonbiomass fuels, higher-income households can afford cleaner fuels. However, it does not explain why government LPG subsidies that have been given out since the late 1960s have not helped poorer households and have benefited only the rich (Srivastava and Rehman 2006; Balachandra 2011a).

Ineffective Policy and Governance for Energy Access
Another reason for poor energy access, especially in India, is related to ineffective federal policies (Bhattacharyya 2006; Balachandra 2011a; Bhattacharyya and Srivastava 2009). Rural electrification during the first two decades after independence typically focused on village-level electrification; this meant that a village could be considered electrified without a single household connection. Even with the establishment of the Rural Electrification Corporation (REC) in 1969, the goal was to boost the rural agricultural economy and thus village-level electrification. It was only with the establishment of the Kutir Jyoti scheme, designed to provide poor households with a single-point lighting connection, that household electrification drew national attention. However, many of these campaigns lacked exclusivity for rural household electrification until the 2003 Electricity Act

and the Rajiv Gandhi Grameen Vidyutikaran Yojana (RGGVY) program in 2005 (Balachandra 2011b).

In addition, there are deep governance problems within the Indian power sector (Modi 2005; Dubash and Rajan 2001). These include problems with electricity generation, transmission, and distribution, as well as the poor financial condition of state electricity boards (SEBs). In 2010, per capita electricity consumption in India was 884 kilowatt-hours (kWh), lagging far behind most industrialized countries and even other developing nations like China (2,658 kWh), Brazil (2,384 kWh), and South Africa (4,803 kWh) (Government of India 2017). This dismal situation is compounded by the fact that India has some of highest transmission and distribution losses in the world. In 2010, these losses were around 24 percent compared to a world average of roughly 10 percent and were trailing other developing countries, including South Africa (11.4 percent), China (6.5 percent), and Brazil (19.6 percent) (Government of India 2017). Moreover, transmission and distribution losses did not come down over the year. In fact, India experienced around 15 percent of losses when it became independent, and the current level is similar to what it was in the early 1990s.

These stark figures show the extent to which the Indian power sector lags behind other countries. Until power sector reforms during the 1990s, the primary responsibility for power distribution lay with the SEBs. However, mismanagement, billing problems, electricity theft, and lack of investment left most SEBs in financial insolvency (Bhattacharya and Patel 2007; Sharma, Nair, and Balasubramanian 2005). While reforms of the power sector have alleviated some problems, especially with the unbundling of generation, transmission, and distribution and with the opening up to private investment, it is still grappling with most of the issues discussed. Studies that focus on inefficient governance in the power sector suggest a technical or regulatory solution whose aim is to reduce the level of transmission and distribution losses (Singh 2006; Thakur et al. 2005; Mathavan 2008). Others focus on the merits of the reforms initiated in the 1990s, and whether liberalization and private investment in the sector will help achieve better outcomes as well as the most efficient organizational structure (Lock 1996; Lal 2006; Bhattacharya and Patel 2007; Dossani 2004).

While these suggested solutions are useful, they ignore the political-economic dimension of the problem. The financial insolvency of the power sector and inefficiency of power generation, transmission, and distribution have their roots in the politics of the country. To better understand how efficient governance of the power sector contributes to rural electrification, we need to look at the larger context of building institutional capacity. This involves not just SEB finances but also an effective bureaucracy and an independent regulatory authority with clear regulations (Dubash and Rao 2008; Dubash and Rajan 2001). It also involves coordination and monitoring across the different ministries responsible for the generation, transmission, and distribution in the power sector and those responsible for expansion of rural electrification initiatives.

Currently, national and state-level power ministries work with mainly public utility companies to manage the expansion of rural electrification. However, this approach results in a focus on selling energy to current customers, while improving the capacity of the grid and using new technologies take a back seat (Balachandra 2011b). Implementing effective policies requires national political parties to have an incentive to pass such legislation. Without a discussion of these incentives, the explanation of ineffective policies is incomplete. Min and Golden (2014) discuss some of the current perverse electoral incentives in which political parties provide free electricity to get votes. While this is a useful starting point, Min and Golden use satellite data as their outcome measure, which could capture street lighting, yet our analysis focuses exclusively on rural household electrification. India's subsidy regime warrants special attention too. The primary aim of these subsidies, initially given to agricultural consumers as an electricity subsidy, was to boost rural economic growth, but they have become a fixture woven into today's Indian politics. During the 1970s, rural farmers realized that these subsidies were very beneficial and began to influence the political process in order to keep them. Soon political parties began campaigning on the provision of free electricity, and the government was spending about a quarter of its budget on these subsidies (Dubash and Rajan 2001). Given that rural farmers were a major voting bloc, they succeeded in ensuring that the subsidy regime would remain in place for decades to come. While these subsidies have had some positive effects in promoting the agricultural economy and increasing rural incomes (Briscoe and Malik 2006), they have also been associated with the financial problems in the electricity sector, making the financial sustainability of these regimes problematic (Monari 2002). Any efforts to reform this subsidy regime have had little success since it is not in the interest of any political party to remove lucrative subsidies (Min and Golden 2014; Badiani, Jessoe, and Plant 2012; Chattopadhyay 2004).

Ineffective national policies have also contributed to India's mediocre cooking energy access situation. The National Project on Biogas Development was among the first national initiatives to distribute biogas cookstoves to the rural population, but these suffered from limited take-up and have mainly favored richer households (Srivastava and Rehman 2006). It was only in 2009 that the government launched the Rajiv Gandhi Gramin LPG Vitrak scheme, a rural LPG distribution scheme, which aims to improve rural LPG access. However, it has yet to give households enough incentives to move away from traditional biomass, leading to fuel stacking: households continue to use biomass cookstoves in addition to LPG (Palit, Bhattacharyya, and Chaurey 2014). Ineffective policies alone cannot explain the dire situation of poor access to cooking fuels. As in the case of rural electrification, this explanation overlooks the political-economic determinants of the problem and the reasons behind why there have been no consistent national-level policies. For example, consider the case of LPG subsidies: since LPG subsidies mainly benefit higher-income households, there is a strong argument to be made for reducing them (Gangopadhyay, Ramaswamia, and Wadhwa 2005; Balachandra 2010), but taking away any existing subsidy regime is a political minefield. In short, the ineffective policies and poor governance explanations do not adequately explain rural energy access in India.

Lack of Decentralized Solutions

Another possible reason for the persistence of energy poverty in India is the lack of decentralized solutions. As far as rural electrification is concerned, these studies suggest that the way forward is to move away from grid extension toward off-grid electricity in order to decouple electricity generation from the national electricity grid (Chaurey, Ranganathan, and Mohanty 2004; Millinger, Møarlind, and Ahlgren 2012; Palit, Sarangi, and Krithika 2014). The primary argument of such studies is that the expansion of grid electricity is slow and has historically favored more populated areas. In such cases, it is better to use off-grid solutions to provide electricity to rural households. An added benefit of such solutions is that it is possible to combine them with the use of renewable sources like water, wind, and solar power. Using such renewable sources means less use of dirty fuels like coal, currently the main source to power India's national grid. This decentralized model, it is suggested, will be more efficient by allowing villagers to set up their own small-scale power generation systems (Hiremath et al. 2009) while leaving the regulation of such off-grid solutions to the national government (Kumar et al. 2009).

Notwithstanding the strengths of off-grid solutions and their associated potential for renewables, they do come with some disadvantages. Typically off-grid power generation is more expensive than grid electricity on a per unit basis, and there is also the possibility of increased costs related to the storage of excess power (Kaundinya, Balachandra, and Ravindranath 2009; Bhattacharyya 2006). Moreover, when there is a decision to expand the national grid to a specific area, this may reduce the incentives for rural villagers to adopt or continue using off-grid systems. A better way to think about off-grid electricity is thus not as a substitute but as a temporary solution integrated with the national grid expansion plans (Urpelainen 2014). This involves boosting institutional capacity through the incorporation of an off-grid component into the larger national electrification strategy. This approach has a number of significant benefits: it reduces uncertainty, allocates capital efficiently across both grid and off-grid initiatives, and facilitates the integration of electrification projects with broader development initiatives in rural areas. In the Indian context, the approach would involve formal coordination among ministries, improving transparency of rural electrification plans, and clarifying regulatory guidelines (Urpelainen 2014). In short, off-grid electricity provision should not be seen as a panacea to the rural electrification problem that India faces. Instead, it would be one component of a broader electrification strategy.

The lack of decentralized solutions does not explain the poor cooking fuel access situation in India. The Indian government has made attempts to distribute improved cookstoves throughout the country since the 1950s, the most prominent of which is the introduction of the National Programme on Improved Chulhas (NPIC) in 1985 (Rehman and Malhotra 2004). Given that the use of biomass was still very prevalent throughout the country, the government determined that the distribution of improved cookstoves that reduce indoor air pollution was a necessary first step before attempting to promote the use of clean fuels. But the main goal of the NPIC was promoting cookstoves with chimneys to reduce indoor smoke and ultimately reduce firewood use. The program did achieve some basic milestones, but it was plagued by problems of high adoption cost and poor take-up rates and ultimately came to a formal end in 2002 (Rehman and Malhotra 2004; World Bank 2002). In other words, the provision of decentralized solutions on its own cannot explain why much of rural India continues to use inefficient biomass stoves for their cooking needs.

Progress and Problems in Rural Electrification at the National Level

Given the limits of the existing explanations, how can we explain temporal and spatial variation in energy poverty in India? We begin by addressing the question of rural electrification in some depth. The first part of the analyses focuses on the national level. Then we consider subnational variation across some Indian states.

The typical qualitative case approach would be to examine policies in India with other similar countries and gain an understanding of what policies work and what policies do not. For example, IEA (2010a) compares rural electrification policies in Brazil, China, India, and South Africa to identify successful initiatives. While this approach can be fruitful to identify potentially useful policies, it suffers from a key limitation that the contexts in these countries are very different. China is an authoritarian country, India is a democracy, Brazil was a military dictatorship from 1964 to 1985, South Africa experienced an apartheid era from 1948 to 1994, and these different historical contexts make cross-country comparisons difficult. Another problem with the typical cross-country analysis is a tendency to select on the dependent variable, that is, to identify potential policies using only successful cases (Barnes 2011). The main problem with this approach is that the focus on successful (or unsuccessful) cases leads to a potential bias on what precise policies work (or do not work). In this chapter, we overcome some of these limitations by applying the subnational comparative method (Przeworski and Teune 1970). This method makes controlled comparisons over time and space within a given country. This approach has the advantage of keeping the context relatively congruent when making comparisons. It explains the uneven nature of development outcomes along both spatial and temporal dimensions (Snyder 2001).

Beginning our analysis with examining changes in rural electrification outcomes at the national level over time offers two primary advantages. First, it allows us to hold the unit of analysis constant and examine whether our theory is valid over different periods. Indeed, rural electrification levels within a country change, and any valid theory would need to explain this temporal variation. The explanatory variables of any theory would also need to vary in order to account for changing rural electrification levels. Second, a within-country longitudinal analysis allows for a clearer exposition of the causal processes at work, making sure that they hold across different levels of both the explanatory and outcome variables (Doner 2009). With an examination of the temporal variation, the national level is the appropriate level of analysis, especially since many of the major electrification policies in India are designed by the central government, which has the incentive and resources to plan for the production, distribution, and transmission of electricity throughout the country. This is not to say that state governments do not matter. In fact, states are critical for the implementation of electrification policies. We discuss decision making at the state level in more detail when we examine spatial variation within the country.

There are also two main advantages of supplementing our analysis with studying the within-case spatial variation. First, it allows us to explain variation in rural electrification levels across different parts of the country. Successful rural electrification policies at the national level do not necessarily translate into similar success across all parts of a country, and examining the subnational level is useful to explain why some parts of the country benefit more than others. Second, it allows us to test whether the theory is consistent across both national and subnational units. In a country like India, states play an important role in implementing rural electrification policies, and a useful theory should also be valid at the subnational level. For our purposes, the subnational comparative method involves examining the progress and problems in rural electrification at the national level in India from 1947 to the present, as well as evaluating rural electrification policies across different Indian states.

An immediate issue when dealing with spatial variation is to determine the appropriate level of analysis. Though rural electrification policies in India have typically had a national flavor, the implementation of these policies relies critically on Indian states. The Electricity Act of 1948 explicitly gave this power to the states with the creation of SEBs (Samanta and Sundaram 1983). Their mandate included the implementation of national regulations in the power sector, expanding the grid and providing reliable electricity to residents of their state. Many of the key decisions made regarding rural electrification have happened at the state level, making it the most appropriate unit of analysis. The second issue when dealing with spatial variation is to determine which states are appropriate for our discussion. We chose five states that have achieved different levels of rural electrification: Uttar Pradesh, West Bengal, Orissa, Gujarat, and Maharashtra. In the rest of this section, we focus on the temporal dimension and leave the discussion of the spatial dimension to the next section.

Research Design

When explaining changes in rural electrification rates at the national level over time, we need to determine the different phases that allow a controlled comparison. One possibility is to partition the period between 1947 and the present based on the rule of different national governments. During this period, the national government in India has been led by different political parties: the Indian National Congress (INC), the Bharatiya Janata Party (BJP), Janata Party, and Janata Dal, with some periods of governance involving national coalitions like the National Democratic Alliance (NDA) and the United Progressive Alliance (UPA). The partition based on the rule of the different national governments is problematic for two reasons. First, the reigns by these political parties vary in length, with some reigns as short as two years and others as long as thirty, making meaningful comparisons of rural electrification policies during these periods infeasible if our goal is to keep contextual factors fairly constant within a given partition. Second, there is substantial variation in rural electrification levels during some of the longer reigns, such as during Indian National Congress Party rule from 1947 to 1977. We would thus miss out on this variation within a given partition if this entire thirty-year period were grouped together in one phase. Another possibility is to split the last sixty-five years based on different levels of rural electrification rates—group periods into bins of low, medium, and high levels of rural electrification to compare changes in policies within these three groups. The main problem with this approach is that the split is based on changes in the dependent variable, leading to selection issues, which we sought to avoid when using the subnational comparative method.

Hence, the partition that we adopt in this section is based on two factors: the different phases have to be (1) temporally congruent and (2) involve significant changes to rural electrification levels in the country. This means that the time needs to cover the period from 1947, when India became independent, to the present, and there should be no gaps between the phases. In other words, the temporal partition that we adopt should be both mutually exclusive and exhaustive. Similarly, the time periods should involve significant changes to the dependent variable, the national rural electrification levels, while keeping most other contextual variables constant. Following the method of difference (Mill 2011), this allows us to control for many of the factors that typically remain constant over time, like geography and type of government, and focus on how changes to our main explanatory variables (government interest, institutional capacity, and local accountability) influence rural electrification levels between different phases.

We measure changes to rural electrification using both the NSS data and the 2001 and 2011 government censuses. The main advantage of these sources is that we are able to measure rural electrification levels at the household level. This is a better measure than other studies that use satellite data to measure electrification rates (Min 2015; Min and Golden 2014) because these are likely to capture street lighting and might thus underreport the level of rural household electrification, our main outcome of interest. The use of these sources does have a disadvantage, especially because our study spans more than seventy years. Reliable data from the NSS are mainly available after 1987, so before this period, we use secondary data, which primarily cover village electrification levels. Reliable and complete census data on household electrification are available only for the two most recent census rounds.

The main challenge in measuring government interest is to avoid circular reasoning. It would be easy to fall into the trap of interpreting positive outcomes as evidence of interest, but such an approach would conflate the dependent and explanatory variables. Therefore, we instead examine the presence and changes in political and economic incentives, independent of the outcome. We then verify that these factors were related to actual investments of resources into improving rural electrification rates so as to provide direct evidence that the government saw an opportunity in rural electrification—again, independent of the outcome. For example, setting up a national-level program that focuses specifically on rural electrification is an indication of a high level of direct government interest. This reflects the motivation of the government to gain at the polls through voters who benefit from such a program. The source of government interest may also be indirect. For instance, the government may increase rural electrification rates by electrifying irrigation pumps, which boosts agricultural productivity in rural areas.

We measure institutional capacity based on the functioning of the SEBs, foreign investment in the electricity sector, and the number of and coordination between national-level ministries to promote rural electrification. If the finances of an SEB are in the black, if there is considerable foreign direct investment in the production, distribution, or transmission of electricity, or if there are clear procedures in place for coordination among different government ministries, then we can say that institutional capacity is high compared to when any of these factors is missing. For instance, while the SEBs in India were running at a loss in the 1980s, the Maharashtra State Electricity Board (MSEB) was the only SEB at the end of that decade that was in the black. This reflects a high level of institutional capacity.

To measure local accountability, we look for mechanisms that allow the rural population to signal their need for electricity to the government and communicate whether they are willing to copay to extend the electricity infrastructure to their area. We also search for instances in which the rural population is able to punish the local political representative for not providing reliable electricity access to their area. For instance, the Communist Party in West Bengal instituted a three-tier system of local government at the district, block, and village levels with direct elections for all three tiers, thereby incentivizing the government bureaucracy to be responsive to the needs of the rural population.

Based on our partitioning of the temporal variation and the measurement of the outcome and explanatory variables, rural electrification in India can be divided into three major phases: (1) 1947 to 1965, when there was low government interest in rural electrification, low institutional capacity in the power sector, and low levels of local accountability for the rural population; (2) 1965 to 2003, when there was a high government interest in rural electrification, a low but increasing institutional capacity in the power sector, and low local accountability for the rural population; and (3) 2003 to the present, when there was continued high government interest in rural electrification and a high institutional capacity in the power sector while local accountability continued to be low. Based on our theory, we should expect to see low levels of rural electrification in phase 1 and medium levels

Table 4.1
Phases of rural electrification in India.

Phase	Years	Government Interest	Institutional Capacity	Local Accountability	Rural Electrification
1	1947–1965	Low	Low	Low	Low
2	1965–2003	High	Low/medium	Low	Medium
3	2003-present	High	Medium/high	Low	Medium

of electrification in phases 2 and 3. Table 4.1 summarizes these phases and the resultant outcome for each phase.

Phase 1: 1947 to 1965

Rural electrification levels in India were very low at the time of independence. Over the next fifteen years, the government made slow progress in expanding rural electrification. Much of the government's focus during this period was the growth of the industrial sector; rural electrification took a back seat. Even in places where the government decided to expand the grid to rural areas, the focus was mainly on village, and not household, electrification. The reason for the increase in village electrification dates back to the Electricity Supply Act of 1948, when the national government decided to create SEBs that were responsible for the generation, transmission, and distribution of electricity. This was part of the Nehru-Mahalanobis strategy, which revolved around raising the per capita income of the country by promoting heavy industry. Dating back to the National Planning Committee of the Congress Party in the late 1930s, this strategy was the brainchild of the future prime minister Jawaharlal Nehru and academic Prasanta Chandra Mahalanobis. It revolved around building infrastructure, machines, and improving productivity output since these were seen as essential parts of kick-starting the postindependence economy. An inherent belief of this strategy was that industrialization would produce capital goods, and the Indian economy would become less reliant on labor. In short, Nehru's vision of the Indian economy was similar to that of the East Asian developmental state and was institutionalized through the Planning Commission and five-year plans (Corbridge 2010).

The First Five-Year Plan (1951–1956) attempted to shift the economic focus from agriculture to industry, leaving rural development and rural electrification on the back burner. Although rural electrification was considered an important tool in stemming the migration of the rural population to the cities, the primary aim of this initiative was to electrify the suburbs of large towns rather than villages. By the mid-1950s, all towns with a population above 50,000 were electrified, and the Indian economy began to show signs of growth. However, more rural parts of the country still received no electricity, with only 1 percent of villages with populations fewer than 5,000 electrified during this period (Samanta and Sundaram 1983, 31). This First Five-Year Plan also sought to bring the states into the production of electricity in order to move away from the British system of power sector governance. By the end of this period, the proportion of electricity produced by public utilities overtook that of private and nonutilities (Kale 2014a).

The goal of the Second Five-Year Plan (1956–1961) was a continued focus on industry, aiming to provide electricity to all towns with populations of more than 10,000 and to 85 percent of towns with populations between 5,000 and 10,000. Also included in this plan was the electrification of 8,600 villages with fewer than 5,000 inhabitants. While this was a useful start to village electrification, there was no return on investment in the rural areas. The primary focus of the plan was still on industrial development and industrial consumers' needs. Industry consumption was already more than 60 percent in the mid-1950s and was expected to increase to more than 70 percent by the end of the period. The agricultural consumption during this period languished at 4 percent (Samanta and Sundaram 1983, 24). The government managed to achieve a large part of what it set out to do in the Second Plan—the number of villages and towns electrified during this period exceeded targets—but rural electrification continued to lag behind.

The ambition of the Third Five-Year Plan (1961–1966) was to reorient policy away from heavy industrialization. However, that did not automatically mean that the rural economy made it to center stage. While the Five-Year Plan noted that electricity was being used for irrigation in rural parts of the country, its goal was more to develop small-scale industry. As part of this initiative, the government aimed to electrify an additional 21,000 villages; in fact, the plan exceeded the target, with around 45,000 villages electrified by 1966. As with earlier plans, although rural electrification was considered an important need, the amount of initial seed capital and the return on investment in rural areas were deemed major barriers. Notwithstanding that the government managed to electrify a number of villages, generating capacity targets fell short of what was planned during this period (Samanta and Sundaram 1983, 25). Part of the story were efforts to reduce the independence of the SEBs, especially in terms of pricing structure. With industrialization and industrial needs as primary concerns in the 1950s, the SEBs were charging industries less than they were agricultural and residential customers (Kale 2014a). The result was that many SEBs were unable to expand their power generation, transmission, and distribution capabilities, and many of them were already in financial trouble.

In sum, at the end of phase 1, rural electrification levels were still very low in the country. By 1963, only Tamil Nadu and Kerala had more than 25 percent village electrification (Kale 2014a). How can we explain the low levels of rural electrification during this phase? Conventional explanations like geography and urban bias are limited in their use. The population of the country was increasing rapidly during this period, especially in rural areas. Moreover, the states of Tamil Nadu and Kerala had no greater population density than other states like Punjab. Therefore, low physical access to the rural population because of geography fails to explain rural electrification levels during this phase. Urban bias provides some explanation of why rural electrification levels remained low compared to cities and industrial areas. However, the focus on small-scale industry during the Third Plan goes against the urban bias reasoning. Small-scale industry in India was focused not so much on urban city centers but smaller towns, and electrification of these areas cannot be accounted for by urban bias alone. Finally, while India's extreme poverty explains the low level of rural electrification to a significant extent, the central and state governments found resources for other expensive pursuits, such as heavy industrialization. For a more complete explanation, we must thus turn to our theory, which focuses on the need for high levels of government interest, institutional capacity, and local accountability.

There was no national initiative for rural electrification during this phase. As noted, Nehru's predominant concern was industrialization. By the middle of the century, India was deemed to become a federal democratic republic, and Nehru envisioned a bigger role for the state in the economy. To this end, the SEBs were given the mandate to improve the electrification infrastructure in their territories, but this also meant that they had the national prerogative to favor industrial consumers. Modernization of the country was the national slogan at the time and was seen as a "diffusion process wherein great pulses of social and economic change—ultimately liberating and uplifting, if often disruptive of established ways of being in the short run—would push outwards from India's major cities to its smallest towns before reaching into the countryside" (Corbridge 2010, 307). While recognizing that electricity was increasingly being used in rural areas for irrigation pumps, this was not a government priority, and no specific national-level program focused on the electrification of India's rural areas.

Institutional capacity during this phase was also very limited. When India was under British colonial rule, many of the electricity production facilities were sited in the major cities and catered primarily to British companies located in these areas. This resulted in village electrification levels around 0.64 percent at the turn of independence (Samanta and Sundaram 1983; Government of India 2017). In other words, there was no colonial basis to improve the electricity infrastructure in the country, unlike, say, the railway network in India, which was fairly extensive during the period of British occupation (Kale 2014a). Given the postindependence national government's focus on modernization through industrialization, there was only a limited increase in the production, transmission, and distribution of electricity in rural areas of the country. The creation of the SEBs was an important step, but the SEBs existed primarily to serve industrial consumers during this phase. The support for industrialization was so high that the SEBs had preferential pricing for industrial consumers compared to agricultural and rural consumers, and some of them were already in financial disarray by the mid-1960s (Kale 2014a).

Many of the electrification policies during this phase were top down. The National Planning Committee of the Congress Party had already decided by the late 1930s on the economic strategy for the country after independence. Following the death of Sardar Patel and the defeat of Purushottam Das Tandon, both senior figures in the Congress Party, Nehru was the dominant voice in the Congress Party by the early 1950s (Corbridge 2010). During this phase, the Congress Party had widespread support through the country, allowing Nehru to put his economic plan into place until his death in May 1964. Although elections were held every five years, there were no mechanisms in place for the rural population to signal their electrification needs to the government. Much of the rural population was uneducated at this time and was not organized enough to participate meaningfully in the political process. Hence, low levels of government interest, lacking institutional capacity, and almost no local accountability provide us with a clear understanding for the extremely low rural electrification levels in India during the 1947–1965 years.

Phase 2: 1965 to 2003

The second phase of rural electrification in India begins around the mid-1960s when the country faced major droughts that affected agricultural production. Instead of the typical five-year plans, the government decided to put in place three annual plans (1966–1969) and sought to shift the economic focus on small-scale irrigation. The food shortage due to the drought forced the government to increase rural, pumpset electrification. Figure 4.4 shows the number of pumpsets that were electrified between 1950 and 1980, and we can clearly see that the rate of pumpset electrification increased significantly after the mid-1960s. During this period, there was renewed government interest in improving the rates of rural electrification. The government set up a Rural Credit Review Committee to extend credit to the agricultural sector, aiming to improve small-scale irrigation through electrified pumpsets. In 1969, it also established the Rural Electrification Corporation with the chief aim to finance rural electrification. The financing was to be done through special rural electrification bonds, the promotion of rural electric cooperatives, and administering any funds received from the central government (Samanta and Sundaram 1983, 26). It began lending to the SEBs and within three years, more than half a million pumpsets and nearly 30,000 villages were electrified. The sharp focus on pumpset electrification had a positive effect on the economy: over the next two decades, manufacturing output increased between 10 and 15 percentage points (Rud 2012), but household electrification was still lagging.

The Fourth Five-Year Plan (1969–1974) targeted improving the generation, transmission, and distribution of electricity and emphasized agricultural development (Samanta and Sundaram 1983, 27). The National Commission on Agriculture, which was appointed during this period to help electrify the rural economy, recommended that fSEBs focus on rural areas for electrification. This plan also exceeded the electrification target in terms of village and pumpset electrification, but electrification in this context was still on tube wells and irrigation pumps to boost the agricultural

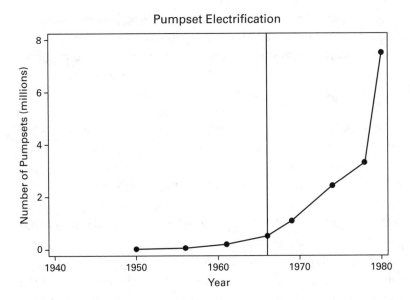

Figure 4.4
Pumpset electrification in India, 1950–1980. The vertical line refers to the beginning
of the second phase of rural electrification. Source: Samanta and Sundaram (1983).

sector; household electrification rates were yet to pick up pace. In the Fifth
Five-Year Plan (1974–1978), the Minimum Needs Programme (MNP) and
rural electrification were integrated (Samanta and Sundaram 1983, 29). The
emphasis now was on places that were below the poverty line, and so a
number of rural areas fell under this plan. This was an important moment
for rural electrification in India, especially since it signaled the central government's efforts to prioritize rural villages that were very poor and had
dire electrification needs. While this reorientation on rural electrification
aided the rural economy and development, the policy outcome was still
measured in terms of the number of villages and pumpsets electrified. This
meant that household electrification was still not a top priority.

The Committee on Power, set up in 1978, was instrumental in shifting
the government's thinking away from agriculture (Samanta and Sundaram
1983, 30). It reviewed the working of the SEBs, central government organizations involved in electrification, as well as the Rural Electrification Programme and concluded that the country's rural electrification initiative,
among other things, would need to improve the electrical power load in
villages. This recommendation hoped to improve village-based, small-scale

industries as well as increase the rates of domestic and street lighting in rural areas. By making the SEBs' rural electrification plan part of the larger Integrated Rural Development Programme, the Committee on Power aimed to merge rural electrification policies to achieve broader socioeconomic goals. During the Sixth Five-Year Plan (1980–1984), rural electrification policies and the Minimum Needs Programme became even more integrated in the effort to boost the rural agricultural economy, especially small-scale irrigation. States like Uttar Pradesh, Bihar, West Bengal, Orissa, and Madhya Pradesh benefited the most during this period from more intensive MNP efforts (Samanta and Sundaram 1983, 31–32). While the focus on the rural economy was useful in boosting government interest for rural electrification, the emphasis of these programs did not directly aim to increase household-level electrification rates.

Toward the end of the 1980s, the government invested significantly in boosting the rural economy. While this allowed for the expansion of the electricity grid to many rural parts of the country, it led to two further problems. First, rural electrification rates were still abysmally low. In 1987, only around 31 percent of the rural population had access to electricity: the increase in agricultural and manufacturing output during the 1970s and 1980s had not translated into household electrification in rural areas. Second, the SEBs were in deep financial trouble. There were two primary reasons for this situation: agricultural subsidies and corruption (Kale 2014a). In its zeal to support growth in the agricultural sector and the rural economy, the government had doled out huge sums of money as agricultural subsidies. This meant that rural farmers were paying way below the market price for electricity access and the SEBs could not recuperate their investment in the grid expansion. Corruption was also a major issue. In the name of oversight, political leaders interfered with the appointment of personnel and the awarding of contracts. Combined with the fact of inefficient SEBs, the ability of the bureaucracy to improve rural electrification was diminished during this period.

The financial disarray of the SEBs, power shortages, and a looming balance-of-payments crisis forced the Indian government to privatize some parts of the electricity sector. They amended the Electricity (Supply) Act of 1948 and the Indian Electricity Act of 1910 to allow private companies to generate and sell electricity. Combined with the opening up of the economy in general, several private companies were eager to participate in the electricity sector. However, since the SEBs still controlled the transmission and distribution of electricity, these companies sold their electricity only to SEBs, and the long-term sustainability of this approach was in question because of SEBs' financial deficits. In order to alleviate some of these concerns, the government offered private companies a number of concessions that have been criticized as going beyond normal investment risk assessments (Ranganathan 1996). Pressure from international donor agencies like the World Bank led the government to further liberalize the electricity sector, with Orissa becoming the first state in the 1990s to unbundle the SEB into two generation, one transmission, and three distribution companies and also establish state-level regulatory bodies, the state electricity regulatory commissions. Many other states followed suit, and the power sector was now more open and market oriented than before. Paired with other national-level regulatory reforms like the Regulatory Commissions Act in 1998, the Availability Based Tariff, and the Indian Electricity Grid Code, there has been a significant improvement and transparency in the way electricity tariffs are set in the country (Singh 2006). These reforms coincided with an increase in rural electrification rates and reached 50 percent at the turn of the century.

Conventional explanations like geography and availability of energy resources fail to explain the changes in rural electrification rates during this phase. Both geography and the availability of resources like coal remained largely the same as in the previous phase. If these explanations were true, then we should see no change to rural electrification rates toward the end of the 1990s. But although the rate of rural electrification was slow to pick up during the 1980s, a series of power sector reforms in the 1990s helped increase the rate to at least 50 percent. Although economic growth facilitated rural electrification, the beginnings of India's progress in rural electrification coincided with a serious food shortage, and one might say that agricultural electrification was the cause, instead of the effect, of India's improved economic performance. To understand the changes in rural electrification levels during this phase, we once again look at government interest, institutional capacity, and local accountability as the main explanatory variables of our theory.

Government interest for rural electrification was high during this period, as the opportunities afforded by pumpset irrigation in agriculture captivated policymakers. A number of initiatives focused specifically on electrifying rural parts of the country. From the setting up of the Rural Electrification Corporation to the integration with the Minimum Needs Programme, rural electrification was high on the government's agenda. The rise in the levels of government interest compared to the first phase was due to the food crisis, itself a result of the drought in the mid-1960s. Until that time, the government favored industrialization and sought to move away from depending on agriculture for economic growth. This was accompanied by the rise of new technologies in irrigation that used electric pumpsets and the provision of electricity to rural farmers that helped boost both agricultural and manufacturing output during these years. However, national initiatives during this period continued to focus on village-level electrification and did not emphasize extension of the grid to rural households. While the rural economy did well, household electrification still lagged behind, especially in the 1970s and 1980s. That said, the high level of government interest was a necessary component to improve rural electrification rates in the country.

Institutional capacity during this phase also improved compared to the previous phase. The emphasis on rural electrification gave the SEBs the mandate to expand the grid to rural parts of the country, which they would not have done otherwise. The grid expansion required substantial investment, and the government was willing to mobilize necessary resources. However, in an effort to incentivize rural farmers to boost their agricultural output, the government decided to provide agricultural subsidies and also meddle with the internal workings of the SEBs. This directly contributed to the dire financial situation of the SEBs at the end of the 1980s. Nonetheless, reforms in the 1990s with the privatization of production, the unbundling of production, transmission, and distribution of electricity, and the setting up of state and central electricity regulatory commissions meant that the financial situation of SEBs improved considerably by the end of this phase.

Local accountability continued to be limited during this period. Much of the rural population was still uneducated and could not mobilize resources to put pressure on the government to signal their electrification needs. During the 1970s, the Congress Party and its leader, Indira Gandhi, embraced a period of rural populism, which was then continued by the Janata Party. However, the effect of such populism was that rich farmers could get subsidies that have remained a political issue (Teitelbaum 2011). Along with the power sector reforms in the 1990s, there have been public hearings to allow people to voice their needs and complaints to the authorities. While this mechanism is useful, it has not been sustained during this phase, and the extent to which the rural population was able to voice their concerns is not clear. In short, this phase witnessed a high level of government interest, low to medium levels of institutional capacity, and continuing low levels of local accountability, and these have resulted in medium levels of rural electrification.

Phase 3: 2003 to the Present

The next phase of rural electrification in India began in 2003 when the central government introduced a new piece of legislation, the Electricity Act of 2003, replacing existing legislation like the Indian Electricity Act, 1910, the Electricity (Supply) Act, 1948, and the Electricity Regulatory Commissions Act, 1998. Its aim was to formalize the opening of the electricity sector to private players for the generation, transmission, and distribution of electricity.

As far as rural electrification was concerned, the act allowed license-free generation and distribution in rural villages, as well as the management of rural distribution by communities at the local level, so called panchayats, cooperative societies, and other local organizations (Singh 2006). The appellate tribunal was set up by the act to allow feedback and appeals regarding the decisions of the central and state electricity regulatory commissions. In addition to the Electricity Act, the National Electrification Policy of 2005 and the Rural Electrification Policy of 2006 allowed central and state governments to work together to draw up plans and provide electricity to rural households through the extension of the grid or other stand-alone systems. Specifically, the Rural Electrification Corporation under the Ministry of Power was given the mandate to coordinate plans with state governments, state utilities, and other government agencies (IEA 2010a).

Along with these regulatory changes, the government also launched the RGGVY scheme in 2005. This program was meant to promote grid extension and was specifically intended to improve rural household electrification levels. As part of this expansion, the RGGVY planned to build a rural electricity distribution backbone that included a substation with sufficient capacity for every block, a village electrification infrastructure that had a distribution transformer, and decentralized distributed generation systems where the extension of the grid was not feasible (IEA 2010a, 67). In addition, the RGGVY advocated the use of renewable energy in these stand-alone systems through the use of diesel generating sets that use biofuels, biomass, solar photovoltaic cells, and small hydropower plants. During the Tenth Five-Year Plan (2002–2007), 235 projects at a cost of 97.33 billion rupees were being implemented, aiming to electrify more than 65,000 unelectrified villages and to provide free electricity to 8.31 million below-poverty-line (BPL) households. The completion rates at the end of this period were lower than expected, but about 39,000 villages were electrified and around 670,000 BPL households had electricity connections to their homes (IEA 2010a, 68). This rural electrification initiative was extended in 2008 and has now continued into the Eleventh Five-Year Plan. The effects of this program have been significant, with rural electrification rates at the end of 2010 at 73 percent, a more than 20 percentage point increase in only a decade.

Another initiative during this phase was the Remote Village Electrification (RVE) program that also began in 2005 under the Ministry of New and Renewable Energies (MNRE), which aspired to electrify villages with fewer than 100 inhabitants, especially those in remote areas and those not electrified by the decentralized distributed generation component of the RGGVY (IEA 2010a). Since this program was under the MNRE, the electrification was to be provided using renewable energy sources. In remote locations, electrification was to be done using solar photovoltaic home lighting systems. Although the RVE is a national-level initiative, it has been implemented by states with a 90 percent subsidy by the MNRE. As of 2013, 10,154 villages and hamlets have been successfully electrified (MNRE 2013). A related scheme was the Jawaharlal Nehru National Solar Mission launched in 2009 as part of the Indian National Action Plan on Climate Change. Though it does not concentrate on the provision of electricity to rural areas, it advocates the use of solar energy for rural electrification (IEA 2010a, 71). Its goal is to have an installed capacity of 20,000 megawatts and the deployment of 20 million solar lighting systems in rural areas by the year 2022.

All of these initiatives reflect a high level of government interest in rural electrification during this phase. The incentive to provide groundwater for irrigation to farmers remained intact, and the electoral incentives to electrify villages grew stronger. Unlike the schemes from earlier phases that focused on village electrification and kick-starting the rural economy, consistent national-level and state-level plans during this phase prioritized the electrification of rural households. Allowing license-free generation and management of electricity distribution by local administrative units like panchayats and cooperative societies in the 2003 Electricity Act are a reflection of how national-level regulations can leverage local resources for implementation. The RGGVY scheme is by far the biggest and most comprehensive national-level initiative by the Indian government to improve household rural electrification rates. It has managed to connect a number of previously nonelectrified villages to the electricity grid and also provided electricity to remote villages where the grid cannot be extended. In short, government interest during this period was very high.

Institutional capacity was also comparatively high. A number of national ministries coordinate policy on power: the Ministry of Power, Bureau of Energy Efficiency, Central Electricity Authority, Central Electricity Regulatory Commission, State Electricity Regulatory Commissions, the Ministry of Coal, the Ministry of Petroleum and Natural Gas, the MNRE, and the Department of Atomic Energy. While the mandates of these ministries overlap to some extent, the Rural Electrification Policy in 2006 envisaged the central and different state governments as coordinating their activities and coming up with a consistent framework to provide electricity to rural households. A good example is the RVE under the MNRE, whose electrification goals were coordinated with that of the RGGVY.

While government interest and institutional capacity were high during more recent years, local accountability still remained low. The government has not sought to put any mechanism in place through which rural villagers could provide feedback about their electrification needs, the reliability of their electricity, and any complaints regarding the affordability of fees. While the extension of the grid through RGGVY has been beneficial to rural electrification levels overall, decisions about which villages to electrify first remain target based rather than needs based; rural populations are largely excluded from these decisions. Access to reliable electricity has made its way into campaign speeches and is now part of the political agenda, but such statements are typically related to free electricity and subsidies rather than attempting to electrify those communities that are in dire need of such access.

Spatial Variation

Now that we have outlined the temporal variation in electricity access for India and also examined how our theory works better than alternative explanations, we are in a position to examine spatial variation across the country. This is an important part of studying electricity access since progress in rural electrification at the national level does not immediately mean improvements to rural populations all over the country. Although India as a whole has implemented plans to improve rural electrification, the benefits of these initiatives have been uneven. Some states like Gujarat, Punjab, and Maharashtra have done well; others like Uttar Pradesh and Orissa have fared less well. Similar to the previous section, we use our theory to study the reasons behind why some parts of India have managed to achieve high rates of rural electrification, whereas others have lagged behind. As explained earlier in this chapter, state governments are responsible for the implementation of rural electrification problems, making it the appropriate level of analysis. We choose five states with different levels of rural electrification and attempt to discern the reasons that these states have had these varied outcomes. The states we discuss are Uttar Pradesh (2011 electrification rate, 24 percent), West Bengal (40 percent), Orissa (36 percent), Gujarat (85 percent), and Maharashtra (74 percent) (Government of India 2011c).

Existing explanations are of limited value to understand this spatial variation in rural electrification rates. For instance, geography does a poor job of explaining the spatial variation in rural electrification. West Bengal is among the most densely populated Indian states with more than 1,000 people per square kilometer, and yet it had achieved a rural electrification rate of only around 40 percent in 2011. Compare this to Gujarat, which has just over 300 people per square kilometer but achieved a rural electrification rate of 85 percent in 2011 (Government of India 2011c). The type of state government is also not useful to understand why some states have done a better job than others. Maharashtra has achieved over 70 percent rural electrification rate, whereas Orissa still languishes at less than 40 percent (Government of India 2011c) even though the Indian National Congress has been the dominant party in both states.

Similarly, the very different electrification levels in West Bengal (40 percent) and Kerala (92 percent) (Government of India 2011c) cannot just be attributed to Communist Party rule in both states. Similarly, resource availability does a poor job of explaining the variation in rural electrification levels given that Orissa is a major coal-producing state but had less than a 40 percent rural electrification rate in 2011. Finally, the explanatory power of per capita income is weak despite its correlation with rural electrification. For example, in both Gujarat and West Bengal, around 22 percent of the population is below the poverty line (Government of India 2013), but Gujarat has a rural electrification rate of 85 percent compared to West Bengal at only 40 percent in 2011 (Government of India 2011c). Overall, existing explanations do a poor job of explaining the spatial variation in electrification rates in the country.

To the best of our knowledge, Kale's (2014a) is the only scholarly analysis that conducts a systematic study of the spatial variation in rural electrification rates. Using Maharashtra, Orissa, and Andhra Pradesh as her cases, she argues that institutional structures that were set up soon after independence have an effect on the rural electrification levels today. Her in-depth analysis is useful for a historical understanding of the challenges facing these states in terms of rural electrification. Indeed, some of the decisions about electrification taken during colonial times have implications for the role of the state in the electricity sector. For instance, Kale (2014b) shows that private companies managed to procure licenses for electricity generation and distribution in the Bombay presidency. This process played a significant role in limiting the ability of the Maharashtrian state from entering the electricity sector (compared to Mysore, where the state was able to invest in hydroelectricity much earlier). These historical events are important in understanding the future government interest of Indian states in expanding rural electrification and improving local accountability of the state government.

While the focus of this section is on the variation in electrification rates among Indian states, we do not claim that all states had the same level of flexibility when it comes to implementing electrification policies. There is a clear link between the national and state levels in India (Singh and Srinivasan 2004): policy design and changes at the national level have implications at the state level. This is especially true in the power sector, where reforms are usually made at the national level (Singh 2006). However, this does not mean that states do not have any autonomy over the implementation of policies. The RGGVY is a good case in point. In 2005, this program began as a national initiative to electrify over 100,000 villages across the country. The Ministry of Power and the Rural Electrification Corporation were the primary drivers behind this effort and were the main funding source (Banerjee et al. 2015), but the implementation of the program was the responsibility of the state governments. They have had autonomy over its prioritization, institutional structure, and deployment speed, and the state electric utilities are typically required to submit detailed implementation plans to the REC before funding is approved (Banerjee et al. 2015). So, while national initiatives are important for improving rural electrification access, states do play a critical role.

To gain better insight into why some states have higher levels of rural electrification than others, we turn to our theory. Recall that we argue that government interest is a necessary component for success in rural electrification. We will show that Orissa is one of the states with low government interest to provide its residents access to electricity and that this has contributed to its low electrification rates despite being the second-largest coal-producing state in India. However, government interest is not sufficient for high rural electrification rates. Instead, progress requires high levels of institutional capacity and local accountability as well. We discuss Uttar Pradesh as an example where government interest is high but institutional capacity and local accountability are low, and these have been at the root of poor rural electrification records. Having either high institutional capacity or high local accountability could improve the probability of success in rural electrification. Good examples of these are Gujarat with high institutional capacity but low local accountability and West Bengal with low institutional capacity but high local accountability.

Finally, according to our argument, government interest combined with high level of institutional capacity and high local accountability will almost certainly ensure success in rural electrification. One state that features all these attributes is Maharashtra. In the rest of this section, we discuss each of these states in turn and demonstrate the merit of our theory in explaining the spatial variation in rural electrification rates within India. The values of the relevant variables are summarized in table 4.2.

Other than secondary sources, we use representative household-level surveys to make our case in this section. These include the Indian government's National Sample Survey (NSS) (NSS 2012a), two rounds of the India Human Development Survey (IHDS) (Desai et al. 2007; Desai, Dubey, and

Table 4.2
Summary of subnational cases.

State	Rural Electricity Access (percent)		Government Interest	Institutional Capacity	Local Accountability
	2001	2011			
Orissa	19.4	35.6	Low	Low	Low
Uttar Pradesh	19.8	23.8	High	Low	Low
West Bengal	20.3	40.3	High	Low	High
Gujarat	72.1	85	High	High	Low
Maharashtra	65.2	73.8	High	High	High

Source: Government of India (2011c).

Vanneman 2012), and data from the 2001 and 2011 Indian government censuses (Government of India 2011c). The NSS data are useful since they allow us to accurately calculate the percentage of rural households with electricity access for the whole country and reliable data are available since the late 1980s. A limitation of these data is that they use nonzero expenditure data for both grid connections and other electricity sources, and so it might overestimate the actual electrification rate. The IHDS survey is valuable because it contains questions about the quality of electricity access at the household level. Specifically, we use a question on the availability of the number of hours of electricity per day among rural households since this measure goes beyond the simple dichotomy of electricity access. We also use primary survey data from a 2014–2015 study by Aklin et al. (2016) conducted in West Bengal, Uttar Pradesh, and Orissa (and three other states not included in our analysis) since it has questions on the household's satisfaction with their electricity supply and with the performance of their state government.

The results are summarized in table 4.2. As the table shows, Orissa and Uttar Pradesh have clearly made less progress in rural electrification compared to West Bengal, Gujarat, and Maharashtra. Orissa and Uttar Pradesh languished below 40 percent of rural electrification in 2011, when Gujarat and Maharashtra achieved electrification rates of more than 70 percent. Only 20 percent of rural households in West Bengal had electricity in 2001, but the state has made considerable progress, doubling its electrification rate in the following decade.

Three patterns in this table deviate from our expectations. First, it is surprising that Uttar Pradesh has performed much worse than Orissa, despite high government interest in the former. As we can see, this lack of performance reflects the very low institutional capacity and almost total absence of local accountability in what is India's largest and among the most backward states. Second, West Bengal's excellent performance stands out despite the lack of high institutional capacity. As the case study shows, this largely reflects the state's rapid improvement in institutional capacity recently, along with an institutional capacity that puts a clear emphasis on local accountability. Finally, Gujarat's performance appears to be better than Maharashtra's, even though the former scores lower in terms of local accountability. Here, the standard control variables are relevant.

Maharashtra is a much larger state than Gujarat, and though Maharashtra's population density initially appears high, almost one in five people in that state live in the Mumbai metropolitan area. Also, rural Gujarat is much wealthier than rural Maharashtra: according to NSS data, in 1999 the average household expenditure in rural Gujarat was 4,780 rupees and only 4,100 rupees in Maharashtra (in 2010 constant prices). A decade later, the difference remained 4,950 rupees to 4,590 rupees in favor of Gujarat.

Orissa

Orissa, located in eastern India, had some of the lowest rural electrification rates in the country in 1987, when only around 17 percent of its rural population had access to electricity (NSS 2012a). More than twenty years later, about 36 percent said that they had access to electricity (Government of India 2011c), but this belies the low government interest as well as major institutional capacity and accountability problems the state faces. Even among rural households that have electricity, power was available for only around sixteen hours per day in 2012, down from nineteen hours five years earlier.[8] Orissa is one of the most economically backward states in India and has a high proportion of indigenous population (about 40 percent belong to the [socially disadvantaged] scheduled caste or scheduled tribe community). During the 1990s, Orissa was among the first that came on board when the national government decided to privatize the electricity sector. However, the reason behind this decision was a focus on industrialization, resulting in continued low rural electrification rates (Kale, 2014a).

Orissa's state government has not shown much interest in improving rural electrification rates. The state has large reserves of coal, bauxite, and iron, and its economy is based on the extraction of these natural resources. As a result, the government has typically favored industrialists, and rural electrification has not been considered a primary development goal. Since agriculture is not critical to the state's economy and the availability of groundwater is mainly in the eastern part of the state near River Mahanadi (Central Ground Water Board 2012), the government has not invested in providing electricity to irrigation pumps as in other Indian states. Much of this dates back to the colonial period, as the coastal areas of the state were governed by the British and were under direct taxation and the inland areas were taxed through intermediary princely states (Kale 2014a). After independence, Congress Party leaders originated from the industrialized coastal areas and have been dominant in Orissa politics ever since (Bailey 1960). In short, the government has had little political incentive to improve rural electrification rates.

Linked closely with low government interest is poor accountability to the rural population. Orissa has a fairly large rural population in dire need of electricity, but since they do not contribute much to the economy, they do not have much influence on the government's policies. There is also considerable social stratification in the state (Ray 1974): the scheduled castes and the scheduled tribes primarily reside in the interior part of the state, whereas the politicians hail from the more urban coastal regions.[9] The rural population did have a party to fight for their interests in the Swatantra Party. However, it faded after 1974 when the national government stopped paying the descendants of the princely states in rural Orissa, the main funding source for the party (Erdman 2007, cited in Kale 2014a). Since this point, the Congress Party and its main rival, the Janata Dal, have governed the state, but both parties have catered to the industrial class, leaving the interests of the rural population out of contention. This has resulted in their favoring industrialization through the mining of its natural resources rather than the development of its agricultural sector. So unlike the farmers in Maharashtra and Punjab, the rural population in Orissa does not have a voice to lobby for electrification.

In addition to poor accountability, levels of institutional capacity were also low in the power sector in Orissa. Though the state emphasized the industrial sector, it did not develop its power sector (Kale 2014a). The Orissa

State Electricity Board (OSEB) was similar to other SEBs in that it faced significant financial problems. It was inefficient and had trouble paying the private companies that generated electricity (Rajan 2000). The World Bank took a special interest in improving the power generation capacity in Orissa by funding the Upper Indravati hydroelectric project (World Bank 1995; Kale 2014a). It was conditional on power sector reform, and the project was delayed for a number of years during the 1980s. In the 1990s, the World Bank decided that it would only fund projects where state governments unbundled their electricity and allowed private players into the electricity sector. The Orissa state government saw this as an opportunity to increase power generation capacity for the industrial sector (Kale 2014a), and in 1996, the OSEB was split into Grid Corporation of Odisha and the Odisha Hydro Power Corporation. The Odisha Electricity Regulatory Commission was also formed at this time. The split of the OSEB was meant to increase efficiency in the electricity sector, but it created a lot of coordination problems between the generation, transmission, and distribution companies, resulting in fairly high transmission and distribution losses (Dubash and Rajan 2001; Rajan 2000). Overall, the privatization program did not succeed in improving the institutional capacity of Orissa's electricity sector to any great extent.

In sum, low government interest has limited any progress for rural electrification in Orissa. Even if the government was incentivized to implement better rural electrification policies, the state suffers from low levels of local accountability and institutional capacity. Rural communities in the state are located in inland areas and are not politically organized enough to influence state policies. The political parties that have governed the state have favored their urban, more affluent constituencies and have also supported the mining sector over agriculture. In addition, the privatization of the OSEB led to the disbandment of the rural electrification division, and the private industry had no incentives to provide electricity to the rural inland indigenous communities.

Uttar Pradesh

Uttar Pradesh, located in northern India, is the country's fourth largest state by area and the largest state by population, with nearly 200 million residents, 78 percent of whom live in rural areas (Government of India 2011a). In 2011, only around 24 percent of its rural population had access to electricity (Government of India 2011c). These rates had not changed much from a decade earlier, when electrification rates were around 20 percent. Even among rural households that report having electricity, power is available for less than ten hours per day on average in 2012, unchanged from five years earlier.[10] Moreover, residents of Uttar Pradesh were more dissatisfied with their electricity compared to those in Orissa and West Bengal in a recent survey (Aklin et al. 2016). What explains this dismal state of affairs in Uttar Pradesh? We argue that the Uttar Pradesh government has had high government interest in rural electrification especially given its untapped groundwater resources. Yet the state has suffered from poor institutional capacity and low local accountability. Both factors have contributed to the reason that the state has lagged behind many other Indian states in providing its rural population access to electricity.

According to the Central Groundwater Board (2012), Uttar Pradesh has the country's largest groundwater resource during the monsoon season. This means that the state government has a clear incentive to put policies in place that regulate this groundwater use and provide electricity to irrigation tube wells in rural parts of the state so that farmers can tap this abundant resource. Since the electoral vote in Uttar Pradesh lies in its rural areas, the government has the necessary impetus to show that it is regulating groundwater use. Since 1970, it has adopted the central government's Model Bill to Regulate and Control the Development and Management of Ground Water. While this policy was fairly sufficient in the initial decades, the Uttar Pradesh government realized that there was overexploitation of groundwater in the state and it needed to update these regulations at the turn of the century. In 2010, the state government proposed the Uttar Pradesh Ground Water Conservation, Protection and Development (Management, Control and Regulation) Act to further regulate the amount of groundwater used for irrigation (Cullet 2012). Specifically, the act sought to establish the Uttar Pradesh Groundwater Authority that would monitor groundwater use and advise the state government. It also allowed for the creation of water user and resident welfare associations that would build capacity at the local level to self-regulate groundwater use. While this bill is yet to be enacted, it reflects the high level of government interest to manage groundwater resource and, hence, extend electricity to rural areas.

Institutional capacity in Uttar Pradesh has always been poor. The Uttar Pradesh State Electricity Board (UPSEB) was established in the late 1950s with the responsibility of providing reliable electricity to its mostly rural population. However, like other state electricity boards in the country, the UPSEB was in deep financial trouble by the end of the 1980s. Around this time, the operating loss of the UPSEB was around US$250 million,[11] and this grew more than fourfold by the end of the 1990s (Gurtoo and Pandey 2001). The main reasons for this low institutional capacity were identified as high transmission and distribution losses, productivity losses, and inefficiencies within the organization and agricultural subsidies. In turn, power sector reform in Uttar Pradesh has followed the standard formula as in the rest of the country with the unbundling of generation, transmission, and distribution, as well as the privatization of generation and distribution utilities. However, as Gurtoo and Pandey (2001) argue, the UPSEB has deeper problems. There are issues with the high cost of power, as well as misreported information on agricultural consumption and the effect of subsidies in the state. Although the unbundling process was followed in Uttar Pradesh, the reform process has been implemented in an ad hoc manner, resulting in continuing poor institutional capacity.

Local accountability, like institutional capacity, has also been poor in Uttar Pradesh. Its rural development is associated with the Community Development Programme put in place in the 1950s and its decentralization of power (Lieten and Srivastava 1999). While the decentralization process allowed more marginal voices to be heard, it remained distinct from the power struggle between the different political parties at the state level. Political mobilization by the lower-caste population through the Bahujan Samaj Party (Duncan 1999; Pai 2002) has largely focused on symbolic politics at the expense of developing institutional mechanisms for local accountability (Jaoul 2006).[12] As a result, much of the existing social hierarchy has remained in place. While land reform policies have allowed greater ownership among the scheduled caste/tribes and other backward classes in the state, the societal structure still continues to be dominated by the middle- and upper-class landed elites. Lieten and Srivastava (1999) also show that institutions that are responsible for greater local accountability, like the panchayati raj, or "local government rule," have been captured by elites, and many of these institutions do not include members of the rural population that is supposed to benefit from them. In other words, local accountability remains very low in Uttar Pradesh, and adequate channels for common people to voice their needs and concerns to the political elite are largely absent (Lerche 2000).

In sum, Uttar Pradesh is a case where government interest in rural electrification is high given its abundance of groundwater resources. Yet the state government has not put in place regulations that would harness this resource and expand the electricity grid to rural areas so that people can benefit from irrigated pumps. Institutional capacity is very poor, with financial losses of the UPSEB mounting during the 1980s and 1990s. Power sector reform in the past decade has helped to some extent, but overall institutional capacity still remains low. Local accountability has always been limited in Uttar Pradesh. State politics continues to be between elites competing for power, and ordinary villagers are left out of the process. These factors have resulted in low rural electrification levels throughout the past seven decades and continue to the present despite government interest being present.

West Bengal

The eastern state of West Bengal until recently has suffered from some of the lowest levels of rural electrification in the country. At the turn of the century, rural electrification rates fell below 20 percent, but the state managed to double its electrification rate within a decade (Government of India 2011c). Moreover, residents of West Bengal also reported higher levels of satisfaction with their electricity supply and with the performance of the state government than Uttar Pradesh and Orissa did in a recent survey (Aklin et al. 2016). What explains the drop in electrification levels in West Bengal during the 1990s and the progress since then? In this section, we argue that the main reason for the initial poor electrification rates was the low level of government interest and institutional capacity. Since then, there has been a marked change in the level of interest, especially in the management of groundwater resources, and this has contributed to a modest increase in rural electrification rates. Local accountability has always been high in West Bengal, especially since the Communist government came to power in the late 1970s. Although institutional capacity has improved modestly over the last decade, it is still low and overall electrification levels remain poor, with nearly 25 million people without access to electricity.

As in the case of Uttar Pradesh, there has generally been sufficient groundwater availability in West Bengal. According to the Central Groundwater

Board (2012), the state has considerable groundwater resources in both monsoon and nonmonsoon seasons. Government interest in regulating this resource is an important gauge of its interest in extending rural electrification because electricity is needed in rural areas for pumping groundwater for irrigation. Given that all political parties in the state compete with each other for rural votes, the government has a clear incentive to exhibit its seriousness in regulating groundwater use. The Communist Party in West Bengal, whose primary constituency lies in the rural areas, has been active in regulating groundwater use, especially since 1993. At the time, a flawed methodology revealed a serious depletion of groundwater in the state (Mukherji 2006), and the state government decided to implement a regulation, the SWID certificate, to control groundwater levels in different areas. While this led to a dip in the number of irrigation pumpsets initially and also contributed to a drop in the rural electrification levels, which fell to around 27 percent toward the end of the century (NSS 2012a), these numbers improved after a revised methodology was employed. A critical aspect of this revised procedure was the incorporation of water table trends to identify whether a particular area had an abundance of groundwater (Mukherji 2006). This change reclassified a number of once-critical areas in West Bengal as safe for irrigation. Although the Communist Party is no longer in power, party competition in the state ensures that the government has an incentive to regulate groundwater levels and extend rural electrification as a consequence.

Institutional capacity has generally been poor in West Bengal, mainly because of the functioning of the West Bengal State Electricity Board (WBSEB). Like many other state electricity boards, until the early 1970s, the WBSEB charged rural farmers based on their consumption, that is, fees were set based on usage through metered tube wells. However, this practice changed in many states, including West Bengal, in the 1980s. The WBSEB moved away from metered tube wells and instituted a flat fee, coinciding with the rise of the Communist Party of India-Marxist (CPI-M) in the state, as well as the ease of administering a flat fee (Mukherji, Shah, and Verma 2010). However, the flat fee was politicized and kept low to appease the rural voter base of the CPI-M. In turn, this meant that there was an overexploitation of groundwater and financial problems for the WBSEB. The CPI-M did not attempt to solve the institutional capacity problem through reform, but just raised the fees, charging West Bengal rural farmers the highest flat fees in the entire country until 2007 (Mukherji, Shah, and Verma 2010). Since then, power sector reform has been implemented fairly well in West Bengal. The WBSEB has gone back to metering of agricultural tube wells, a move supported by the rural farmers (Mukherji and Das 2014). This change has also boosted the finances of the WBSEB and modestly improved the overall institutional capacity of the state.

Local accountability has always been high in West Bengal. Though the farmer lobby has not been as strong as in states like Maharashtra, the CPI-M government that ruled West Bengal for nearly four decades was generally responsive to the rural population. When the CPI-M came into power in 1977, it implemented a three-tier system of local government at the district, block, and village levels (Bardhan and Mookherjee 2006b). It also instituted a system of direct elections for all three tiers, ensuring that the government bureaucracy was responsive to the needs of the rural population. Reforms made to the system in the 1980s and 1990s focused on creating a bottom-up budgeting system (Bardhan and Mookherjee 2006b). This system of local governments has been very effective: it has led to higher political participation and similar benefits for the scheduled caste/scheduled tribe groups compared to the rest of the village population (Bardhan et al. 2009). While this has led to clientelistic relationships between the Communist Party in the state and some rural parts of the country, the level of local accountability has remained high in West Bengal.

In conclusion, West Bengal has exhibited high government interest, especially in managing farmers' groundwater use. The methodological error that made the state government institute barriers for tube well licenses has been relaxed since the late 1990s. West Bengal also has high levels of local accountability. Although the CPI-M government may have instituted direct elections in the state to weaken the strongholds of their rival Congress Party, it allowed the state government structure to be responsive to the needs of the rural population. Where West Bengal has lagged behind is the institutional capacity of its power sector. The financial situation of the WBSEB was poor for most of the 1980s and 1990s because of the flat fee system. However, the state has implemented power sector reform over the past decade and has made most improvements to its institutional capacity. We argue that this is what underlies recent improvements in terms of rural electrification in West Bengal.

Gujarat

Gujarat, located in western India, has a population of 60 million (Government of India 2011d) with an economy dependent on both an agricultural and an industrial base. The farmlands in the center and southern part of the state consume around 45 percent of the electricity output, and its industrial sector, which houses around 30 percent of India's chemical and textile industry and 12 percent of the country's sugar industry, uses around 28 percent of its electricity output (Hansen and Bower 2003). At the end of the 1980s, rural electrification levels in the state were at 43 percent, but this increased to almost 80 percent within a decade and went up further to 85 percent by 2010 (NSS 2012a; Government of India 2011c). The quality of electricity access is also high in Gujarat: among rural households that report having electricity, power was available for more than twenty-two hours a day in 2012.[13] These massive increases in rural electrification rates can be attributed to high government interest in rural electrification, especially after Narendra Modi became chief minister in 2001. It is also an example of high institutional capacity, with the Gujarat State Electricity Board (GSEB) resolving its financial difficulties after power sector reforms. Agricultural lobbyists have usually influenced local policies, but this meant that only rich farmers saw the benefits of rural electrification, and so local accountability is on the relatively lower side.

Government interest in improving rural electrification rates has always been high in Gujarat. The government's incentive stemmed from the fact that the agricultural sector has been a substantial component of the state's economy and that rural farmers form a critical voting constituency for all political parties. The importance of the agricultural sector is particularly acute in Gujarat given that groundwater is a scarce resource. During the monsoon season, groundwater availability increases at a rate comparable to that of Orissa and Andhra Pradesh, but recharge from rainfall is almost zero when there is no monsoon (Central Ground Water Board 2012). Since the agricultural sector consumes a large share of electricity, management of groundwater resources is an important factor in determining the extent to which a government is interested in putting in place policies that favor rural electrification (Mukherji 2006). Since gaining independence, the Gujarat government has been fairly responsive to the needs of the agricultural sector. The demand to increase agricultural output in the state was the primary driver in building new pumps as well as extending transmission and distribution lines to different parts of the state. Government interest received a boost after the election of Modi as chief minister. He put in place a number of policies that directly improved rural electrification rates. The most important of them was the Jyotigram Yojana project, which provided different electricity feeders to rural areas. That meant that nonfarm rural customers of the state would be able to get a dedicated power supply, one that was not interrupted by the needs of the industrial sector. Launched in 2003, this scheme has tremendously improved the electrification rates, as well as the quality of electricity, in rural Gujarat over the past decade. One study has claimed that there were almost no power cuts and voltage fluctuations in Gujarat within five years after the start of this program (Shah and Verma 2008).

Together with government interest, institutional capacity has also been high in Gujarat. Agricultural demand had the effect of increasing the state's institutional capacity to provide electricity to different areas. However, this also meant that there was initially a strain on the GSEB. The subsidies provided to the agricultural sector implied that the Gujarat SEB did not have the capacity to generate sufficient power, and it had to buy additional power from independent power producers (Dholakia 2002). This in turn led to a deteriorating financial situation. The establishment of the Gujarat Electricity Regulatory Commission improved the financial situation of the GSEB, and the state electricity board was also broken down into separate production, transmission, and distribution units. This unbundling helped reduce a lot of the inefficiency and corruption surrounding the GSEB, and politicians were replaced with bureaucrats to run these organizations. Thus, institutional capacity was restored in the state after reforms in the power sector.

Local accountability is on the lower side in Gujarat. As with other Indian states like Maharashtra, the agricultural sector had been powerful and vocal in Gujarat politics up until 2003. Given that the level and quality of surface water is insufficient for their needs, farmers have historically influenced state policy to focus on the electrification of pumpsets. That meant that village-level electrification was favored over the provision of electricity to rural households. In fact, the Central Electricity Authority has declared that Gujarat had been 100 percent electrified since 1988. However, household-level electrification still hovered around 80 percent in 2005. Rural household-level electrification suffered because there were local communities that did not have the means to communicate their electricity needs to the state government. In 2002, households with higher levels of income reported much higher levels of electrification: in fact, those with a yearly income of less than 20,000 rupees had only around 60 percent electrification (Hansen and Bower 2003). Part of the problem can be attributed to the infrastructure cost and the focus on the agricultural sector. However, a major problem was the top-down nature of much of Gujarat's electrification policies, and "local councils do not have adequate [authority] to bypass the central government's rural electrification program and provide energy services directly to households" (Hansen and Bower 2003, 18). While local accountability improved in the decade after the election of the Modi government, a lot more remains to be done to institutionalize local accountability like in West Bengal.

Gujarat is a case of high government interest, especially in the new millennium. The management of groundwater resources in water-scarce Gujarat was an important initiative by the state government. Institutional capacity in the state was low during the 1980s when its financial situation was dire, but the management of the sector improved after the unbundling of the GSEB. Local accountability in Gujarat has come in the form of agricultural farmers' influencing state policies toward rural electrification. While this is similar to other Indian states like Maharashtra, it has favored village level over household electrification. Because of this, poorer rural households had been excluded from the rural electrification bandwagon. Local accountability is relatively low in Gujarat but has improved in the past decade with measures that target specifically household-level electrification in rural areas.

Maharashtra

Maharashtra in western India has had much higher rates of rural electrification than Orissa, Uttar Pradesh, and West Bengal. Although the situation in the state was bleak in the late 1980s, with just 37 percent rural electrification (NSS 2012a), the next two decades saw rates rise considerably to 74 percent in 2011 (Government of India 2011c). The reason for this increase can be linked to high government interest, high institutional capacity, and high local accountability. Maharashtra is endowed with a high level of groundwater resources, making agriculture an important part of the state economy, which incentivizes the government to invest in the electrification of irrigation pumps. The Maharashtra State Electricity Board (MSEB) faced the same pressures as other SEBs, but it managed its financial situation much better than others did. The state also has a strong farmer lobby that influenced rural electrification policy, leading to higher levels of local accountability (Kale 2014a). Though the focus was mainly on electrifying farms and agricultural production, there was a spillover into rural household electrification. When market reforms were introduced throughout the country, these local groups were also able to resist privatization and ensure that their rural interests were protected.

The first two decades after independence witnessed a reorganization of political power in Maharashtra (Baviskar 1980). While its economy continued to be driven by the urban industrial sector based out of Bombay (present-day Mumbai), political power shifted more to rural areas of the state. The main reason for this shift was the rise of the Maratha community, a rural class of mostly chieftains, farmers, and landowners, in the state's political arena (Datye 1987). They accounted for nearly 40 percent of the state's population and also formed the core constituency of the Congress Party. It was precisely this electoral incentive that shaped the interest of the Congress Party in investment in the rural economy and allowed it to remain dominant until the mid-1990s (Palshikar and Deshpande 1999). The Congress government made a determined choice to build the agricultural sector, thereby ensuring that the rural Maratha farmers (known as the "sugar barons") would benefit the most during these times (Lalvani 2008; Kale 2014a; Baviskar 1980). The increased investments in the rural economy formed the basis through which the state made progress in electrification, as the expansion of the agricultural economy led to an associated increase in rural household electricity access. Even after the mid-1990s with the rise of the Shiv Sena and Bharatiya Janata (BJP) parties (Vicziany 2002), there was high government interest in greater rural electrification access since all political parties keenly contested for the rural Maratha vote.

The MSEB was established in 1960, and its mandate was primarily about rural electrification. Yet it did not have means to generate sufficient revenue to increase rural electrification because the main profit-generating urban center, Bombay, continued to remain in private hands (Kale 2014a). While the rest of the country had public sector electrification before the 1990s and moved to privatize after, Maharashtra has steadily maintained the split model (Das and Parikh 2000), so the MSEB's main source of funding was from state coffers. This money was used to boost production through the use of hydroelectric and thermal power and also increase the number of pumpsets for rural agrarian development (Kale 2014a). During the first fifteen years, farmers were the main beneficiaries of the MSEB initiatives. However, this rapid expansion plan for the rural sector strained the institutional capacity of the MSEB (Das and Parikh 2000). There were massive shortages, and the government had to cease operation of state-run industry and textile mills in 1979 to conserve electricity. The MSEB responded by instituting a flat-rate system of payment. Though there was considerable pressure on the MSEB to change its pricing policy to cater to rural industrialists, it was the only SEB at the end of the 1980s that turned a profit (Kale 2014a). During market reforms in the 1990s, the Maharashtra government decided to pursue foreign direct investment in electricity production. This approach was not successful, as exemplified by the Dabhol Power Plant, an international joint venture of Enron, GE, and Bechtel, which came under pressure over Enron's bankruptcy, but privatization of the MSEB did come through in the early 2000s (Dubash and Rajan 2001). Although institutional capacity in Maharashtra has faced some ups and downs, it has remained fairly high relative to other Indian states.

Local accountability has also been high in Maharashtra. The rise of "sugar barons" meant that rural farmers who controlled vast amounts of land were able to influence state politics (Kale 2014a). They were able to resist vast reductions in state subsidies to their agricultural sector and helped to keep rural electrification on the government's agenda (Lalvani 2008). When market reforms were introduced in the 1990s, the farmers ensured that the interests of the rural population were not exploited through a steep increase in prices. They pressured the government to promote the use of biofuels from sugarcane production waste (Kale 2014a). This strategy was mainly a response to growing pressure from the domestic sugar industry, which was facing fierce competition from other Indian states. With help from the US Agency for International Development, the state launched bagasse plants, which use residue from sugar cane production to produce electricity from biomass. After technical know-how was transferred (Kale 2014a), there was a series of state-led initiatives that provided incentives to improve electricity production through bagasse (Purohita and Michaelowa 2007). Overall, the introduction of bagasse as an alternative to ramp up energy production was generally successful. These efforts ensured that the electrification of rural areas in Maharashtra were always at high levels.

Overall, high levels of government interest in rural electrification as well as high levels of institutional capacity and local accountability increased levels of rural electrification in the state. Although institutional capacity fluctuated over time, the MSEB was profitable at the end of the 1980s compared to its counterparts. By the end of the 1990s, rural electrification rates had increased modestly to over 70 percent. After power sector reforms in the 1990s and early 2000s, rural electrification levels continued to increase in Maharashtra, and by 2011 it had reached around 74 percent (Government of India 2011c). This case exemplifies how high levels of government interest, institutional capacity, and local accountability work together for success in rural electrification.

Progress and Problems in Modern Cooking

Similar to the way in which we examined the Indian government's policies on rural electrification, we now evaluate its policies to provide the rural population with access to clean cooking fuels. We analyze cooking policies separately from electricity because the Indian government's approach to providing access to clean cooking fuels has largely been independent of rural electricity expansion. Moreover, less than 0.5 percent of households that have electricity use it for cooking (NSS 2012a). The government has also not paid as much attention to the provision of cooking fuels as it has to electricity, especially for the rural population (Palit, Bhattacharyya, and Chaurey 2014). The result is that most households in rural areas still use solid, biomass fuels like wood, plant residue, and dung for their cooking needs.

Indian households use LPG and biomass as their two main cooking fuels. Since biomass is proven to increase indoor air pollution and LPG is the cleaner alternative of the two fuels, we evaluate the government's policies on efforts to incentivize households to move away from the use of biomass fuels and encourage adoption of LPG as a cooking fuel, especially in rural households. Overall, government efforts have largely failed. Although the use of LPG increased somewhat over the past few decades, the use of biomass did not decrease significantly throughout the country. To understand the reasons behind the government's failure on both counts, we will again make use of our argument presented in chapter 3.

Recall that based on our theory, strong government interest is necessary for improved energy access. We first study whether national-level initiatives that focus on moving away from biomass use and encourage LPG adoption are strong and sustainable schemes. We argue that the Indian government has made some attempts to reduce biomass fuel use in rural households over the past three decades, but these efforts have largely failed because they have not had a strong, sustainable initiative at the national level. In fact, the efforts have focused not on reducing biomass use but decreasing the level of indoor air pollution through improved biomass cookstoves. Moreover, adoption rates of these improved cookstoves have been poor despite large government subsidies. Attempts to improve LPG access in rural areas have only recently gained national-level attention and still remain nascent at best. Much of the increase in LPG use over the past few decades has been driven primarily by richer households, which have benefited the most from the government's LPG subsidies (Srivastava and Rehman 2006; Balachandra 2010). Hence the government's efforts to steer households away from solid fuels toward the healthier LPG alternative have not been strong enough to be successful. We do not claim that strong national initiatives are sufficient for expanding access to clean cooking fuels, but it is a necessary component if the government wants to make progress on this front.

It is, of course, possible that rural households do not see the benefit of using cleaner cooking fuels, either because they do not understand the health benefits of using LPG or are unable to generate income with the time savings they achieve with using cleaner fuels (Palit, Bhattacharyya, and Chaurey 2014). However, recent evidence suggests that improved access to LPG will increase average household energy expenditures even after controlling for household income in both rural and urban areas (Alkon, Harish, and Urpelainen 2016), and so there does not seem to be any shortage of demand for clean cooking fuels. Despite this, much of the Indian government's effort has been to promote the use of improved biomass cookstoves that reduce the amount of indoor air pollution. The main problem with this approach is that it attempts to provide a technical solution to the issue at hand, and such solutions will achieve only small reductions in biomass usage, leading to a *"chulha* trap" (Smith and Sagar 2014).

The earliest attempts to promote the use of improved cookstoves were in the 1950s, albeit they were very limited. It was only with the introduction of the National Programme on Improved Chulhas (NPIC) in 1985 that a national initiative aimed at distributing improved cookstoves throughout the country (Rehman and Malhotra 2004). Given that the use of biomass was still very prevalent across India, the Indian government determined that the distribution of improved cookstoves that reduce indoor air pollution was a necessary first step before attempting to promote the use of clean fuels like LPG. The program achieved some milestones: improved cookstoves were distributed to some 33 million households, and studies confirmed fuel savings, reduction in smoke levels, cooking time, and time spent collecting fuel (World Bank 2002). However, the government interest for this program was temporary and it came to a formal end in 2002. Problems with the program included a reliance on subsidies, which made it unsustainable; flaws in the design of the cookstoves; ineffective promotion of health benefits; and high adoption costs (Rehman and Malhotra 2004; World Bank 2002). Moreover, the expected health benefits of using improved biomass cookstoves may have been overestimated. Hanna, Duflo, and Greenstone (2016) find that reductions in smoke inhalation do not sustain after the first year of adoption and have no improvements to lung functions. They also found that many of the households did not use the stoves regularly, and rates even declined over time.

After the end of the NPIC program, the Indian government launched the National Biomass Cookstove Initiative in 2009 to expand the use of improved biomass cookstoves. The focus of the initiative was the "setting up of state-of-the-art testing, certification and monitoring facilities and strengthening R&D programs [and the] aim was to design and develop the most efficient, cost effective, durable and easy to use device" (MNRE 2009). This was a slight improvement over the earlier NPIC initiative since it aimed to develop cookstoves that were comparable to clean energy sources like LPG (Venkataraman et al. 2010). The initial focus was on identifying a technological solution to the problem by designing an improved biomass cookstove and setting emission standards and test protocols. As part of the Twelfth Five-Year Plan, the NBCI launched the Unnat Chulha Abhiyan program to help develop and deploy improved biomass cookstoves. The target users of these cookstoves were kitchens of midday meals, *aangwadis* (government-run shelters), forest rest houses, tribal hostels, and small business establishments, as well as households in rural areas. It set an ambitious target of distributing almost 6 million cookstoves by the end of 2017, with 2.5 million going to rural households (India Ministry of New and Renewable Energy 2014). But a recent assessment of the scheme says that it is not well organized and has not received similar interest like the RGGVY (Palit, Bhattacharyya, and Chaurey 2014).

Overall, there has been very little government interest for better access to modern cooking fuels in India over the past seventy years. The NPIC was in effect for almost two decades and did not achieve great success with improved biomass cookstoves. The NBCI initiative has also embarked on distributing better-quality biomass cookstoves throughout all of India. While these are useful interim solutions to reduce indoor air pollution, little government interest has existed to promote cleaner cooking fuels. Government subsidies for cleaner fuels like LPG have mainly favored rich households (Srivastava and Rehman 2006). The Rajiv Gandhi Gramin LPG Vitrak scheme that began in 2009 has improved rural LPG access but has still not given households enough incentives to move away from biomass, leading to fuel stacking where households continue to use biomass cookstoves in addition to LPG (Cheng and Urpelainen 2014). Hence, pushing ahead with national-level initiatives for cleaner cooking fuel access might be the best way forward.

The Future of Energy Access in India

India faces steep challenges in both the provision of electricity and clean cooking fuels to its rural population. There are still 250 million people without electricity access, and more than 90 percent of them live in rural parts of the country. Some states in India, like Maharashtra and Gujarat, have made massive strides in improving rural electricity access, but others, like Uttar Pradesh and Bihar, have lagged behind. By studying the reasons behind the temporal and spatial variation in the country, we established that the best explanation across the board of the current state of affairs draws on government interest, institutional capacity, and local accountability. While government interest has improved over time and is high at the national level today, improvements can still be made in the institutional capacity of the power sector. However, with only a few exceptions, local accountability is still low throughout the country. If India is to move ahead and provide millions of households with access to electricity, the nation needs to make more investments in improving its institutional capacity and local accountability. Specifically, the country will benefit if it improves the efficiency and governance structures of its power sector in production, distribution, and transmission. It will also help if mechanisms are put in place so that rural populations can signal their need for electricity and willingness to pay for such services. This will allow villages that have greater interest and demand in energy to be served before others.

Access to modern cooking fuels in India is an even greater challenge than electricity access. Around 90 percent of the rural population continues to use biomass for their cooking needs. While there was some progress in providing LPG access, these advances have mainly benefited urban customers. By examining the reasons behind the poor state of affairs when it comes to providing LPG access, we identified the main reason as low government interest. Simply put, there is currently no political incentive for the government to improve access to clean cooking fuels. Unlike rural electrification, LPG access in rural areas is not achieved as a by-product of policies to raise agricultural productivity. Because the government is not under pressure to improve access to clean cooking fuels for economic reasons, limitations of institutional capacity and local accountability in this sector have yet to become binding constraints. For advocates of improved access to modern cooking fuels, then, the primary emphasis should be on strengthening government interest. In the context of democratic politics, this means making access to LPG and other alternatives to traditional biomass an electoral issue. Greater awareness of the adverse health consequences of traditional biomass and the availability of cleaner alternatives could result in greater bottom-up demand for more effective policies and programs. Such demand could contribute to improvements in institutional capacity and the creation of local accountability mechanisms.

Our review of the history of energy access in India also suggests some insights for the role of decentralized energy solutions. Technologies such as off-grid solar power and improved biomass cookstoves have become important parts of the energy access tool kit, and in India, their potential has drawn increased attention (Palit, Bhattacharyya, and Chaurey 2014). The history of energy access in India suggests that for new technologies to result in improvements, there must be strong government interest, institutional capacity for policy implementation, and local accountability mechanisms.

In rural electrification, government interest has been strong and increasing over time. Institutional capacity has also increased, although off-grid applications create somewhat different implementation challenges. In this sector, local accountability remains the primary challenge, a factor that could be particularly important for off-grid applications. If India is to use new technology to improve access to modern cooking fuels, the first imperative is to create and sustain government interest. The reduced cost and increased convenience of modern cooking, thanks to new technologies such as cookstoves that mitigate the negative effects of traditional biomass, may remove barriers to sustained, effective government programs. Yet the fundamental challenge is to ensure that the government is interested in improved outcomes in the first place.

5   Country Case Studies: Determinants of Success

The previous chapter showed that our theory can account for India's efforts to alleviate energy poverty. However, it remains to be shown that the causal mechanisms uncovered in the case of India can account for a larger number of cases. This and the next chapter focus on achieving that goal with eleven shorter case studies.

Our sampling strategy seeks to provide a selection of diverse countries across regions and values of variables related to the traditional explanations for energy poverty: geography, resources, and wealth. The cases are summarized in table 5.1. For each country, we report the values of variables that capture geography, energy resources, and income in the year 1980. The table offers two useful observations. First, we confirm the diversity of the sample. This and the next chapter cover a wide range of types of countries. Second, traditional explanations are not sufficient to explain outcomes. South Africa is a prime example: it was comparably rich, it had a lot of natural resources such as coal, and there was no major geographic obstacle to electrification or the provision of modern cooking fuels. And yet energy poverty remained a chronic problem until 1994. In contrast, China, a vast and, until recently, extremely poor country, has made rapid progress in reducing energy poverty over seven decades. On the opposite side of the spectrum, an observer relying on traditional explanations should have been fairly optimistic in 1980 about countries such as Senegal or Kenya. They were wealthier than China, Vietnam, and Ghana, and yet they failed to make progress.

What explains success? To structure the discussion, we begin by describing commonalities across cases of success in this chapter. We consider a case a success if we see considerable, perhaps even unexpected, improvement in rural electrification or cooking. We consider success in either area sufficient

Table 5.1
Summary of cases.

Case	Success: Electricity	Success: Cooking	Population density	Rural population (percent)	Resources	Annual per capita income	Region
China	Yes	Yes	105	81	Coal (M)	$220	East Asia
Vietnam	Yes	No	165	81	Coal (M)	$263	East Asia
South Africa	After 1994	After 1994	23	52	Coal (H)	$5,569	Sub-Saharan Africa
Ghana	Partial success	No	47	69	—[a]	$412	Sub-Saharan Africa
Brazil	Yes	Yes	15	35	Hydro (H)	$4,217	Latin America
Chile	Yes	No	15	19	Oil, gas, coal, hydro (M)	$3,362	Latin America
Bangladesh	Partial failure	No	634	85	—	$244	South Asia
Indonesia	Partial failure	No	80	78	Oil (H), gas (M)	$556	East Asia
Nigeria	No	No	81	78	Oil (H)	$841	Sub-Saharan Africa
Kenya	No	No	29	84	—	$537	Sub-Saharan Africa
Senegal	No	No	29	64	—	$712	Sub-Saharan Africa

This table summarizes the outcomes for each country in chapters 5 and 6. The second column refers to electrification and the third to cooking. Columns 4 to 7 provide data on the traditional explanations for energy poverty: geography, resources, and wealth. Geography is measured as population density, with higher values indicating ease of energy access. It is also measured as the share of the population living in rural areas. Resources are measured as follows. We classify each country based on oil, gas, and coal production in 1980, as well as hydroelectric capacity. If a country produced less than one barrel of oil per person per year, we classify it as low; if it produced between one and five barrels, we classify it as medium (M); if it produced more than five barrels, we classify it as high (H). If a country produced between 1,000 and 5,000 cubic feet of gas per capita, we classify it as medium; anything higher is high. We classify a country as medium if it produced between 0.1 and 1.0 short tons of coal per capita, and high if it produces more. Finally, we classify a country as medium if it has more than 0.05 but less than 0.2 kilowatts of installed hydroelectric capacity per capita, and as high if it produces more. Wealth is measured as GDP per capita in 1980 (1984 for Vietnam). The data are from the World Bank's World Development Indicators, except for hydroelectric capacity which comes from the US Energy Information Administration. Region is based on the World Bank classification.

[a] In the case of Ghana, considerable hydroelectric potential was later developed.

because our theory predicts that progress should be much more common in rural electrification than in access to cooking fuels. The four cases that have achieved considerable success are China, Vietnam, South Africa, and Ghana. With the partial exception of South Africa, all four countries were among the poorest in the world at the time they began making major strides in rural electrification. All four have an authoritarian background, but Ghana and South Africa underwent a democratic transition, whereas China and Vietnam remain firmly authoritarian. Two of the countries are in sub-Saharan Africa and two in Asia, the two regions of the world that have suffered from energy poverty for the longest periods of time. This combination of commonalities and differences allows us to examine the ability of our posited explanatory variables and causal mechanisms to explain varying progress in the eradication of energy poverty over time. China and Vietnam also illustrate the pitfalls of narrowly focusing on democratic institutions in the study of government interest.

Besides these four countries, we report results from Brazil and Chile, where the challenges have been quite different. In a pattern common to most of Latin America, both Brazil and Chile have enjoyed high levels of overall electrification and at least some degree of access to clean cooking fuels for a long time. In both countries, however, the reality in vast rural areas has been different, with much lower levels of access. The democratic transitions of both Brazil and Chile in the last years of the Cold War allow us to explore how middle-income countries can succeed in the end game of eradicating energy poverty. These cases show that our theory can also be applied fruitfully to cases in which the preconditions for success are present.

In the next chapter, we describe cases of partial (Bangladesh, Indonesia) or complete failure (Nigeria, Kenya). Again, we have cases from both sub-Saharan Africa and Asia with mostly low wealth levels, though Indonesia is much wealthier than Bangladesh. Nigeria, depending on one's perspective, is either blessed or cursed with exceptional fossil fuel endowments. We also examine the case of Senegal, in which we see surprising and early progress in the provision of modern cooking fuels without corresponding progress in rural electrification. Toward the end of these two chapters, we offer a summary assessment of these eleven cases and contrast them to our more detailed case of India.

Asian Success Stories: China and Vietnam

In Asia, China and Vietnam have achieved impressive progress in rural electrification and, to some extent, the provision of clean cooking fuels. These
achievements began well before their economic booms and reflected a combination of government interest, institutional capacity, and strong local
accountability mechanisms. The two cases are also interesting because both
have achieved success under authoritarian rule.

China: Seven Decades of Success

China is both the world's most populated country and one of the great
success stories in the eradication of energy poverty. Although one initially
suspects that this success is a consequence of China's phenomenal economic growth after Deng Xiaoping's liberalizing reforms that began in
the late 1970s, this intuition turns out to be wrong. China had achieved
great success with rural electrification during the early years of Mao's rule.
Furthermore, China's success with improving access to modern cooking fuels and cookstoves began well before the country lifted itself out of
extreme poverty through industrialization. These successes can be attributed to a high level of government interest combined with institutional
capacity, especially in later years, and high levels of local accountability
throughout.

China's success with rural electrification began in 1949 when Beijing formulated its first five-year plan (Peng and Pan 2006, 75). At the time, China
was one of the world's poorest countries and had virtually no electricity
access. Moreover, it was not yet in a position to enact a comprehensive
national policy for electrification, so "provincial and local governments
were in charge of the management of power projects" (Bhattacharyya and
Ohiare 2012, 678). By 1958, China had made only limited progress, but
even this progress is impressive against the backdrop of the country's desperate resource scarcity at the time. After Mao's failed Great Leap Forward
between 1958 and 1961, the internationally isolated China made large
investments in self-reliance, including the construction of a thousand
hydropower stations (Bhattacharyya and Ohiare 2012, 678). These stations were localized, however, and did not form a national electric grid. As
the 1970s approached, the central government also began to promote the
extension of the national grid. Already, China's achievements were nothing

short of astounding. By 1978, rural China had achieved an electrification rate of 61 percent (Bhattacharyya and Ohiare 2012, 679).

During the market reforms that began in 1978 with the Open Door policy, China's rural electrification policy became even more dynamic. The central government became increasingly active in rural electrification and used new taxes on power projects to channel revenue into rural electricity generation (Zhao 2001). According to Peng and Pan (2006, 76), "By 1978, under the planning economy mode, China had established a comprehensive management network vertically from the central level, through regions, provinces, prefectures and cities to counties." At the end of the 1980s, these new investments had extended electricity access to 78 percent of China's rural population—a 17 percentage point increase in a decade (Bhattacharyya and Ohiare 2012, 679). For all practical purposes, universal electrification was achieved in 1997, with 97 percent of the rural population having electricity access (Bhattacharyya and Ohiare 2012, 679). Since 1998, China's focus has been on improvements in technical performance, cost-effectiveness, and access to larger loads of power (Peng and Pan 2006, 81).

China is an interesting case also for the provision of modern cooking fuels. While its achievements in this field are less impressive than in the field of rural electrification, they are nonetheless notable (Zhang et al. 2009, 2815). Here, the transition began much later than rural electrification. China's rural energy consumption mix was 71 percent firewood and straw in 1979, and this computation is not limited to cooking. The only notable alternative to these traditional biomass sources was coal, which accounted for 18 percent. By 2007, the share of straw and firewood had decreased to 31 percent, but this comparison is somewhat misleading because it hides a large increase in commercial energy use. Moreover, the main replacement for straw and firewood has been coal, the share of which increased to 43 percent during this time, again with the caveat that some part of this use is for commercial purposes. LPG remains at only 6 percent, suggesting that this modern cooking fuel has not made a breakthrough in China. According to Zhang et al. (2009, 2815), the limited use of LPG reflects "inaccessibility and high price."

Although these numbers do not initially appear impressive, they hide two important facets of China's rural energy policy. First, China can claim credit for the world's most successful program in the promotion of efficient cookstoves (Smith et al. 1993; Sinton et al. 2004). Between 1983 and 1990, the central government and provincial authorities collaborated to distribute a staggering 120 million cookstoves to rural Chinese households (Sinton et al. 2004, 37). In fact, the program was so successful that in 1990, the government decided that the goals had been achieved and moved away from distributing inexpensive or free stoves toward increased commercialization and technological improvement.

Second, China's increased use of coal for rural energy has been a strategic decision, not an accident, and the Chinese government has achieved some success in promoting this goal (Zhang et al. 2009). While coal is often considered an inferior fuel to LPG or electricity for cooking (Sinton et al. 2004, 41), it is inexpensive in China and a clear improvement over traditional biomass in terms of environmental impact (Smith et al. 1993, 949), indoor air pollution (Mestl et al. 2007, 24), and convenience (Smith et al. 1993, 950). From the early 1950s, the central government has encouraged rural households to use coal. This market grew rapidly during the reforms of the Open Door policy, as the central government supported the formation of local coal markets.

Overall, China has been highly successful in mitigating energy poverty, especially through rural electrification. To understand why China has been so successful in mitigating energy poverty, we first consider the fundamentals of energy resources, geography, and economic wealth. While China certainly enjoys an advantage in terms of hydroelectric and coal resources, the existence of these resources cannot explain the government's policy development or decisions such as the formation of a centralized grid. If anything, the availability of local, readily available resources should reduce the need for central policy. But geography is also not a good explanation for success because China's early rural electrification reached the remote and sparsely populated inland. Finally, wealth is a particularly poor explanation for China's achievements, as progress began at a time of abject poverty, well ahead of the liberalizing reforms that have made it an economic powerhouse.

Because these control factors cannot provide a satisfactory explanation for China's success, we turn to our theory. To begin, the evidence is clear on the Chinese government's strong interest throughout the period of investigation, along with the drivers of that interest. While rural electrification has been the priority for China, the government's investments in modern cooking fuels are also notable.

In the case of rural electrification, the origins of the Chinese government's interest varied over the period of investigation, but the level remained high throughout. As Peng and Pan (2006, 83) summarize, "The main driving force of rural electrification in China is a top-down one that relates to national macroeconomic and strategic concerns." As early as 1953, an "administrative agency of Small Hydropower was established ... under the Ministry of Agriculture and training was organized to provide experts" around the country (Pan et al. 2006, 14). At the time, the high level of government interest stemmed largely from China's isolation and the resulting need to emphasize indigenous development in cities and the countryside. Somewhat paradoxically, the failure of the Great Leap Forward in the 1960s was itself a significant impetus to rural electrification, as the central government realized that it must strengthen the rural economy to avoid future hardship and possible political instability (Pan et al. 2006, 15).

In the reform era, the government's interest in rural electrification became more directly connected to the imperative of industrialization in the countryside. Bhattacharyya and Ohiare (2012, 679) note that the drive toward commercialization and industrialization in the countryside led the government in the early 1980s "to move towards an integrated energy strategy and rural energy management." The government's commitment to using economic growth to retain power and secure the support of the rural majority required supplying electricity for both commercial and household uses.

In the case of biomass, a combination of deforestation and public health concerns drove increased government interest over time. China's interests in efficient cookstoves and the use of coal reflect concerns about deforestation and the scarcity of firewood in many areas of the country (Sinton et al. 2004; Zhang et al. 2009). Chinese policymakers had already recognized these issues in the 1950s and have since made concentrated efforts to promote alternatives to traditional biomass. These efforts have not been as ambitious as the Chinese rural electrification program, reflecting the notion that the direct benefits of electrification are much greater for the central government.

If anything, the role of institutional capacity in China's energy access policy has been even clearer than that of government interest. Here we see a pattern of rapidly growing institutional capacity over time. When the Communist Party came into power in 1949, China had virtually no capacity for rural electrification, but beginning as early as 1953, the country developed a set of institutions that were able to guide rural electrification. In 1963, the government announced an emphasis on a top-down approach to centralized grid extension; by 1978, "under the planning economy mode, China had established a comprehensive management network vertically from the central level, through regions, provinces, prefectures and cities to counties" (Peng and Pan 2006, 75–76). For Bhattacharyya and Ohiare (2012, 685), the importance of implementation capacity is one of the key lessons from China for other countries:

The emphasis on training and capacity building, standardisation and dissemination has helped in spreading the knowledge widely across the country. The development of a cadre of skilled technicians and project staff and the performance improvement through feedback loops were also essential factors.

During the Open Door policy, China never allowed its institutional capacity in rural electrification to grow weaker. Instead, Chinese policymakers continued to strengthen its institutions and policies, especially in regard to enabling local communities to become more effective in rural electrification (Peng and Pan 2006, 78).

China's success in the cooking sector has depended on considerable institutional capacity. While evaluating the role of institutional capacity is somewhat difficult in the case of coal market formation due to the lack of data, the success of the efficient cookstove program provides an abundance of information. As Sinton et al. (2004, 48) put it, the "success was based on strong administrative, technical, and outreach competence and resources situated at the local level, motivated by sustained national-level attention." While efficient cookstove programs have proven challenging in many countries, China's Ministry of Agriculture was able to sign contracts with 860 counties for funding (Sinton et al. 2004, 37). In the provinces and counties, rural energy offices were populated by tens of thousands of ministry experts, and Sinton et al. (2004, 38) report survey results indicating excellent administrative capacity, competence, and professionalism across these offices. According to Smith et al. (1993, 943–945), Chinese officials also implemented a series of exacting pilots in a small number of counties to test and improve their intervention.

In contrast to the temporal variation in the availability and relevance of institutional capacity, local accountability has been critical throughout. While Mao's China cannot be said to have featured high levels of account-ability at the national level, the governing committees and councils of the Communist Party of China in the villages and townships of rural areas constantly faced popular pressure to improve rural electrification. At the very beginning, under almost impossibly difficult conditions, China's rural communities made their own investment in small-scale hydroelectric facili-ties and began rural electrification well ahead of other countries at the same level of development (Bhattacharyya and Ohiare 2012, 678). An important reason for this success was an early decision to share the cost of rural elec-trification among multiple levels of government:

> To mobilize resources, the central government adopted an equal share arrangement for investments by the central government, county government and commune/village. Investment from the central and county governments was in the form of capital for equipment and technology, while investment by the rural commune/village was in the form of labor. The dividends received by the central and local government were earmarked for a fund that sought to advance further hydropower development. (Pan et al. 2006, 15)

This logic of burden sharing meant that the central and county govern-ments depended on the contributions of the local population, who ben-efited from government funding to develop local hydroelectric facilities.

Although the central government's institutional capacity strengthened over time, the role of local communities remained strong throughout. According to Bhattacharyya and Ohiare (2012, 679), in the era of economic reforms, the "tremendous progress in terms of rural electrification can be attributed to the new wave of thinking and enterprise reforms carried out in China during this era, which was based on the decentralisation of opera-tions and devolution of powers amongst component units of government to fast-track the electrification process." If anything, local authority appears to have been strengthened as the share of local investment in rural electri-fication increased toward 1990 (Bhattacharyya and Ohiare 2012, 679) and "the central government handed over the administration of rural electrifi-cation to local governments. Power supply bureaus at towns were partially or fully local government institutions" (Peng and Pan 2006, 78). Consider again Bhattacharyya and Ohiare (2012, 684):

Unlike other developing countries that followed a top-down approach to electrification, China has relied on a bottom-up approach, where the local level administration and participation was responsible for the local solution. The approach allowed flexibility and was anchored in self-reliance. Although it may be argued that this started not as a deliberate policy innovation as such in a politically isolated country in its initial days but as a desperate, last resort option of some sort, the credit still goes to the country for retaining this decentralised approach in an otherwise planned, command-oriented economy.

This statement is particularly notable as it suggests that local accountability was by no means inevitable. The Chinese political system was in many ways rigid and hierarchical, with extreme limitations on individual innovation and independent community development, yet the challenges of rural electrification resulted in a historical evolution that allowed local accountability to bloom. Initially the central government's attitude was one of benign neglect: in the years following the revolution, Mao's regime simply did not have the capacity to influence policies on the ground. But over time, the Chinese government never chose to assert its supremacy over local rural electrification institutions. Instead, various cabinets chose to support efforts to enhance local accountability and allow the local population to exert an influence on policy design through relatively autonomous village and township committees. Indeed, the *World Energy Outlook* describes the role of local accountability in China's success story in clear terms: "A key factor in China's successful electrification programme was the central government's determination and its ability to mobilise contributions at the local level" (IEA 2002, 374).

Local accountability also played a key role in the success of the efficient cookstove program. From the very beginning, provinces and counties invested more money into the program than the central government did (Sinton et al. 2004, 39). In 1983, the investment by the central government was 8.2 million yuan, whereas the investment by provinces and counties was 25.1 million yuan. In 1990, when the aggressive expansion phase ended, the investments were 1.3 million and 65 million yuan—showing that the entirety of the program was funded from local funds. According to Smith et al. (1993, 949), households participating in the cookstove program ascribed great value to the technology: "A common sight we found in the households surveyed was a beautiful [improved biomass stove] as a centerpiece to the kitchen, often finished with white tile and kept spotless."

They also note that the Chinese program of piloting was based on careful scrutiny of candidate counties:

The criteria for being chosen as a pilot county go well beyond such obviously salient factors as evidence of biofuel shortages. Equally important are the managerial, financial, technical, and raw material resources existent in the county. Finally, the county must demonstrate the desire to become a pilot county by entering first into a provincial competition ... and then successfully winning in the final national competition.

As this strategy shows, the Chinese program relied on a bottom-up approach that ensured local interest and held local officials to a high standard of performance. The local population's interest in the technology ensured continuous monitoring and provided the local administration with incentives to succeed.

To summarize, China is an ideal illustration of our theory. Strong government interest in rural electrification and, at times, modern cooking fuels laid the foundation for progress. In combination with robust local accountability mechanisms and rapidly growing state capacity over time, this interest has allowed China to claim the position of a forerunner in the global effort to make energy poverty disappear. No other large, developing country has achieved such consistent and impressive success over a period spanning seven decades.

Vietnam: Universal Electrification after Two Decades of Warfare

Vietnam's efforts to eradicate energy poverty began in earnest only after the end of the war in 1975. Over the next four decades, Vietnam achieved virtually universal electrification despite the total destruction of the infrastructure during the war and very high rates of poverty, rural and urban. Progress in modern cooking has been less impressive, and the reason is that the government's commitment to providing modern fuels has been, for reasons of political and economic interest, limited. Similar to China, government interest has varied between rural electrification and cooking, whereas institutional capacity and local accountability have been high, and growing, throughout.

According to Gencer et al. (2011, xi), Vietnam's progress in rural electrification can be divided into four periods. At the end of the Vietnam War in 1975, the country's rural electrification situation was desperate. As Khandker, Barnes, and Samad (2013) write, Vietnamese electric utilities estimate that the rural electrification rate was as low as 2.5 percent. The destruction of the country's infrastructure limited electrification to urban centers, and between 1975 and 1985, the rural electrification rate increased only to 9.3 percent (Gencer et al. 2011, 7). At this time, most rural electricity supply came from isolated systems, but their reach was so limited that Vietnam was barely ahead of the least electrified societies of the world, such as the poor and badly governed dictatorships of sub-Saharan Africa. In fact, Vietnam's success at this time is even less impressive when one considers that it can be largely attributed to recovery from two decades of total warfare everywhere in the country.

In the second stage, however, Vietnam began to prepare for massive rural electrification (Gencer et al. 2011, 8–9). In 1986, Vietnam initiated economic modernization through the Doi Moi policy that brought major liberalizing reforms to Vietnam's economy. While Doi Moi, which began in the aftermath of a serious economic crisis (Van Arkadie and Mallon 2004, 66), itself did not enable rural electrification, it provided rural households with greater incomes due to successful agricultural reform and also created new financial structures that enabled household loans for rural electrification. Vietnam's industrialization drive required large investments in the power sector, creating the generation capacity that would underpin a rural electrification effort. Moreover, the transmission and distribution systems were considerably strengthened, again creating the technical foundations for rapid rural electrification in years to come. But rural electrification of households itself remained limited, as most of the electricity supply was limited to productive uses, such as irrigation. In 1990, the rural electrification rate was still only 13.9 percent.

After this preparatory period, Vietnam began to record truly spectacular growth rates in rural electrification around 1994. Gencer et al. (2011, 11–19) maintain that by 1997, the rural electrification rate had reached 61 percent, making Vietnam's program perhaps the world's fastest: in only seven years, almost half of the country's rural population gained access to electricity. A combination of major transmission investments, especially the construction of a north–south line that created a unified national grid, and a series of policies to promote household connections in rural communities allowed Vietnam to become the world's envy in rural electrification success. Indeed, Vietnam has achieved virtually universal rural electrification today. According to our standard data source (IEA 2017), Vietnam's rural electrification rate reached 98 percent in 2016 and is, thanks to rapid economic growth, still growing.

Vietnam's achievements in the provision of modern cooking fuels are less impressive. According to Heltberg (2004, 880), in 1998, 67 percent of all—not only rural—Vietnamese households continued to rely on traditional biomass only; a mere 10.9 percent had fully left solid fuels behind. While he does not provide the data for the rural–urban difference, it is clear that most of the households climbing the "energy ladder" (Masera, Saatkamp, and Kammen 2000) live in urban areas. According to IEA (2017), 41 percent of all Vietnamese households continued to lack access to modern cooking fuels in 2015. While Vietnam is doing better than India in this regard, where 64 percent lack access to modern cooking fuels according to the same source, it bears remembering that Vietnam and India are fairly comparable in terms of GDP per capita and level of urbanization, at least in 2015. Vietnam is hence doing much better in terms of rural electrification, but the difference in improvements in access to modern cooking fuels is marginal.

Indeed, we could not find evidence of national cooking fuel programs on a large scale. The increase of fuels such as LPG appears to reflect the operation of markets at a time of rapid economic growth without strong government support, as there is no evidence of major government programs during the past decades. The same goes for efficient cookstoves. Although a large number of Vietnamese households continue to rely on traditional biomass, an evaluation of the national market for cookstove technologies found that to this day, "there has not been much interest in the funding of end-to-end, full-scale [improved cookstove] programs by either the Government or NGOs" (GACC 2012, 4).

What explains these patterns? The fundamentals of energy poverty offer only limited insight into Vietnam's successes and failures. First, there is no reason to expect that the fundamentals would have differential effects on rural electrification and access to modern cooking fuels. Moreover, the fundamentals cannot provide a compelling explanation for rural electrification either. While the country is densely populated, the terrain is difficult to penetrate—as American soldiers quickly learned during the war—and the infrastructure was in very poor shape after the war. Before the year 2000, Vietnam produced very little energy itself, though more recently it has been exploiting oil, gas, and coal resources with vigor (British Petroleum

2013). While Vietnam's economic growth has been impressive, rural electrification began at a time when rural poverty was still widespread, and household electricity connections put a heavy burden on ordinary people's incomes. Similar to China, Vietnam achieved rural electrification at a relatively early stage of economic reform, and scholars such as Khandker, Barnes, and Samad (2013) have shown that successful electrification has been an important driver of growing rural household incomes.

Given that the existing explanations are of limited use, we must turn to the political economy. In Vietnam, the government's interest in rural electrification was strong and stemmed from the need to secure regime legitimacy through agricultural and industrial growth for an improved economy. In their comprehensive analysis of Vietnam's rural electrification experience, Gencer et al. (2011, xvii) offer a concise evaluation:

Vietnam's success can be credited to the unwavering national commitment to rural electrification. … Once rural electrification targets were set, and pledges to support rural electrification were made, policy makers stood by them and never backtracked from what was originally promised.

Khandker, Barnes, and Samad (2013, 661–663) similarly state that in Vietnam, rural electrification was a "national priority. … Rural electrification has been a critical component of the government's program to eliminate poverty, redress imbalances in development, and improve overall welfare levels by providing reliable lighting sources, better living conditions, health care, and other rural services."

The government interest story shares some features with China, yet there are also notable differences. The main similarity is the importance of rural electrification as an economic growth strategy. While Vietnam is an authoritarian regime, the legitimacy of the Communist Party has depended on economic prosperity since the end of the war. In this regard, China and Vietnam both exemplify the possibility that a competent authoritarian may choose economic growth as a strategy of political survival, in tune with political economy theories that emphasize the importance of resource extraction for rulers whose support depends on relatively large domestic political coalitions, such as Communist Party members (Bueno de Mesquita et al. 2003).

In turn, a key difference from China is that Vietnam's rural electrification boom only truly began once the central government began making public investments. In China, Mao's regime in the early years was able to support rural electrification through benign neglect, as local communities engaged in electrification through the exploitation of local energy resources, but this strategy seems to have been less successful in Vietnam. Indeed, it was only around 1994, when a national electrification program was already in full swing, that Vietnam began to see excellent results.

But government interest alone would not have allowed rapid rural electrification without equally impressive institutional capacity. In the case of Vietnam, such institutional capacity was developed over time. At the end of the war, institutional capacity in the energy sector was obviously limited, as the government's focus was on warfare. In the coming decades, the government established a hierarchical structure for rural electrification, relying on provincial and regional utilities and their branches for the rural electrification effort (Khandker, Barnes, and Samad 2013, 662). At the time, only planning was conducted at the national level. While this structure was a marked improvement over the anarchic situation in 1975, it prevented the government from making a major coordinated push for rural electrification across the country. By the time of Vietnam's rural electrification growth in the early 1990s, however, Vietnam had developed a capable power sector administration that was able to enact and implement rural electrification policy swiftly and effectively. In January 1995, the entire power sector was consolidated under the holding company Electricity of Vietnam (Asian Institute of Technology 2004, 19). In this structure, a rural electrification agency was included, significantly improving the government's implementation capacity. Statistically, improved capacity can also be seen in sharp decreases in transmission and distribution losses: between 1995 and 2000, these losses decreased from 22 to 14 percent, a stark contrast to the increase from 18 to 22 percent between 1980 and 1995.

Institutional capacity has played an important role in Vietnam's success. As Gencer et al. (2011, xvii) argue,

The policy and regulatory measures introduced by the government, equipping [Electricity of Vietnam] with the mandate and resources it needed to perform its leadership role in a commercially sustainable way, were critical components of Vietnam's success in rural electrification. [Electricity of Vietnam's] emergence as a strong champion for rural electrification in the late 1990s was an important factor for ensuring the technical quality of the rural energy networks and sustainability of rural electricity supply going forward.

Improvements in institutional capacity allowed the government to signifi-cantly scale up rural electrification over time by enabling truly national policy with plans, targets, and priorities. Moreover, as Gencer et al. (3011) suggest, the improved institutional quality of Vietnam's power sector has ensured progress in rural electrification not only in terms of connections but also of the quality of supply. In this sense, Vietnam again shares with China an important feature: while the roots of institutional capacity are to be found in the Communist Party's administrative competencies, the endogenous improvement of the governance of the power sector and rural electrification policy are also important factors that this case study captures.

In Vietnam, the authoritarian institutions at the national level did not prevent high levels of local accountability that have played an important role in rural electrification. Even the slogan of the Vietnamese national rural electrification, "State and People, Central and Local, Working Together," is quite telling (Gencer et al., 2011). According to a case study by the Asian Institute of Technology (2004, 21), the entire Vietnamese rural electrifica-tion distribution system is based on the delegation of the last-mile distri-bution responsibility to local units. The Vietnamese rural electrification program allows considerable flexibility in local management, encourages local contributions, and brings together various interested shareholders. An evaluation study of Vietnam's rural electrification by Crousillat, Hamilton, and Pedro (2010, 88) notes that the willingness of local communities to contribute to the program was one of the cornerstones of its success. While this is not direct evidence for local accountability mechanisms, the strong interest of the local communities at both the household and leadership levels is consistent with the idea that local people were able to channel their preferences and information to policymakers at higher levels and hold them accountable for good outcomes. After all, the national government specifically made local organizations and agencies responsible for the out-comes of the rural electrification, strengthening the accountability chain between officials and people.

Underneath this successful use of local accountability mechanisms is the more exogenous force of strong local demand at the community level:

Local authorities' responsiveness to the strong societal demand, and their choice to accord adequate priority to this issue, and the culmination of this into a national agenda item were critical factors for success. There was persistent dedication and collaboration between central government policy makers and provincial, district, and commune level authorities, as well as EVN [Electricity of Vietnam] and local communities. Once rural electrification targets were set, and pledges to support rural electrification were made, policy makers stood by them and never backtracked from what was originally promised. (Gencer et al. 2011, xvii)

Gencer et al. (2011, 12) provide anecdotal but telling evidence of this demand. They note that in many rural communities in Vietnam, people greeted each other by inquiring about electricity access. When rural electrification works began, local communities organized ceremonies and celebrations that lasted for many days. Gencer et al. (2011, 12) also note that households had to pay at least a full month's income for a connection, emphasizing how much emphasis rural households put on electrification as a life improvement.

These achievements in rural electrification are impressive, whereas Vietnam's success in the provision of modern cooking fuels has been much more limited and recent. The reason, we argue, is limited government interest. In contrast to rural electrification, the provision of modern cooking fuels has not promised direct, tangible benefits to Vietnam's industrial and agricultural growth strategy. While the government has sustained a consistent effort of rural electrification over the past four decades, as we have seen, there has been no such effort to replace traditional biomass and solid fuels with modern alternatives. The issue has not been high on the government's agenda, and therefore little progress has been made. To the extent that Vietnam's recent growth of LPG use can be considered a success, it has been driven by private investment and marketing enabled by growing household incomes, as opposed to government policy. In this regard, the basic association between household income and access to modern cooking fuels seems to be the simplest explanation for this pattern.

Given the lack of government interest, there is not much to be said about institutional capacity or local accountability. In contrast to the rural electrification program, Vietnam does not have an institutional structure dedicated to the provision of modern cooking fuels. There are also no local accountability mechanisms to speak of, as the lack of a national institutional structure means that such mechanisms would not be able to feed into the policy formulation process. We have also found no evidence of widespread grassroots efforts to promote technologies such as modern cookstoves, again in stark contrast to the early efforts of local communities to aggregate and invest funds for power generation and electricity connections.

To summarize, Vietnam presents the contrast between rural electrification and provision of modern cooking fuels in the starkest possible terms. Due to the government's strong commitment to rural electrification for self-interested economic reasons, the combination of growing institutional capacity and strong local accountability mechanisms has enabled rapid progress, to the benefit of the rural energy-poor. In the case of modern cooking fuels, the lack of government interest has prevented such success.

African Success Stories: South Africa and Ghana

With the exception of the Arab countries of the continent's northern parts, Africa is badly behind all other continents in the eradication of energy poverty. The cases of South Africa and Ghana show that when a democratic transition strengthens government interest in reducing energy poverty, there is nonetheless reason for hope and optimism. In both cases, progress in rural electrification has been impressive, and, partly due to fortuitous circumstances, the provision of clean cooking fuel has also improved. In particular, the South African case shows that when the price of electricity is very low, thanks to available energy resources and large past investments in generating capacity, rural electrification can also offer a solution to the problem with cooking fuels. Besides government interest, institutional capacity and local accountability have made important contributions. In both cases, the rapid democratic transition allows us to conduct an informative before-and-after comparison by tracing changes in policy formulation, implementation, and outcomes over time.

South Africa: Energy Access after Apartheid

The most striking feature of the South African case is the dramatic difference between the apartheid and postapartheid era. As soon as the national government's political survival began to depend on the votes of the black majority, the country saw rapid progress in rural electrification. South Africa had inherited impressive institutional capacity from the days of apartheid autarky, and the design of the rural electrification policy after Nelson Mandela's inauguration in 1994 strategically tapped into powerful local accountability mechanisms. Because South Africa's electricity monopoly, Eskom, had historically made large investments in generating capacity, electricity prices were even sufficiently cheap to allow the use of electricity for cooking by the poor. As a result, South Africa's recent success in rural electrification has also enabled notable progress in the provision of modern cooking fuels.

During South Africa's apartheid era, the black majority saw very little progress in terms of electricity access despite the country's wealth and the clear need for electricity in the strategically critical mining sector. As Bekker et al. (2008, 3125) note:

A disparity was also seen in access to basic services and infrastructure, including electricity. The 1996 census, the first census in South Africa that surveyed the whole population, indicated only 58% of the country's population had access to electricity, and only one in four non-urban black South African households was electrified, as opposed to 97% of non-urban white households.

It was only at the end of the apartheid era that the country's democratization created a major institutional opening for a new rural electrification policy, as President Mandela's government now depended on the black majority's votes for political survival.

South Africa's progress in rural electrification was impressive. Largely thanks to the efforts of the state monopoly for electricity, Eskom, the country's total electrification rate climbed from slightly below 65 percent to above 89 percent by 2012, of which less than 3 percentage points did not rely on the national electric grid (SSA 2013, 21). What is more, the urban–rural gap was surprisingly small. While 92 percent of urban households were electrified, the same percentage for rural households was 82 percent (SSA 2013, 22). The contrast to the 1993 number among rural households, 12 percent (DME 2001, iii), is striking. This achievement is not entirely recent either; Bekker et al. (2008, 3130) note that the 70 percent threshold in total electrification was reached early, by the year 2000.

What is more, South Africa's success in electrification also directly contributed to solving the problem of access to modern cooking fuels. While electricity is rarely used for cooking in most developing and emerging economies, South Africa is a singular exception to this pattern. This fortuitous connection between rural electrification and access to cooking fuels stems from unusually low electricity prices relative to household wealth. Between 2002 and 2012, the share of South African households relying on electricity as their primary cooking fuel increased from 58 to 75 percent, while the share of households using wood decreased from 19 to 12 percent. Kerosene use decreased from 16 to 8 percent, and the use of gas increased from 2 to

3 percent only (SSA 2013, 58). These numbers show that in the context of South Africa, a solution to rural electrification is a solution to the problem of gaining access to modern cooking fuels. Indeed, even among rural households, electricity is already the primary cooking fuel, with a 54 percent share in 2012 (SSA 2013, 60). Indeed, Barnes et al. (2009, 6) specifically attribute the rising use of electricity in cooking to the rural electrification policy.

Can the fundamentals of resources, geography, and economic development explain these outcomes? Energy resources, and coal in particular, certainly favor South Africa's rural electrification, but they have been available for a century, and very little was done to eradicate rural energy poverty until 1994. South Africa is a large country with vast, sparsely populated rural areas. Finally, South Africa's status as the continent's economic powerhouse was confirmed well before the end of the apartheid, yet the government made no progress in reducing energy poverty. Therefore, turning to our political economy explanation is more productive.

The South African case is almost a natural experiment in the role of government interest. During apartheid, the government had virtually no interest in rural electrification. If anything, its interest was in preventing rural electrification to ensure that the black majority would remain marginalized and unable to organize politically. However, after the African National Congress (ANC) party won the elections in a landslide, the government's political survival hinged on the votes of the rural majority. Indeed, there is direct evidence that the very formulation of rural electrification was designed by the government in a fashion that minimized the influence of the old apartheid establishment. According to Bekker et al. (2008, 3130),

From a policy and an institutional point of view, the period from 1994 to 1999 may be regarded as a transitional period, during which apartheid frameworks and policies were dismantled or reformed, a new constitution was adopted, new government institutions were created at national, regional and local levels, and other institutional reforms were carried out in many areas of government.

This quote well captures the notion that the government deliberately chose to break with the past and reform the approach to rural electrification for better and faster results. According to Gaunt (2005, 1310), the Electricity-for-All program of the government "commenced that South Africa started going through political change from apartheid to a broadly democratic government." Bekker et al. (2008, 3136) also emphasize the importance of universal service delivery in turning rural electrification into an "impera-tive" for the government.

Quantitative evidence for these claims comes from research by Kroth, Larcinese, and Wehner (2016). They find that South Africa's rural electrifi-cation achievement in the postapartheid period is the largest in communi-ties that had the largest number of previously disenfranchised black people. They also find evidence that Eskom prioritized areas with strong ANC sup-port, again consistent with the idea that democratization creates political incentives for reducing energy poverty.

The strong government interest prompted by democratization allowed South Africa to capitalize on a high level of preexisting institutional capac-ity. To cite Bekker et al. (2008, 3126) again, "The electricity sector faced few of the usual barriers to electrification in developing countries, viz. lack of access to capital, lack of skills and lack of supply infrastructure. In addition, much of the initial demand was for urban electrification, a process far less costly and labour intensive than rural electrification." During the apart-heid era, Eskom had made huge investments in capacity, allowing South Africa to enjoy exceptionally low electricity prices. Fortuitously, Eskom was reformed in the 1980s after economically costly fee hikes:

It slowed down its capacity expansion programme, having accumulated significant surplus capacity. Over time, these investments were amortised so that by the late 1980s and early 1990s Eskom was in a good position, both in terms of capacity and a healthy balance sheet, to launch an ambitious electrification programme. Through the restructuring, Eskom was freed from its previous prohibition on making a profit or a loss, which allowed it crucial leeway in determining the viability threshold for electrification projects. These changes created a political opportunity for Eskom to establish itself as a national champion in the eyes of the new government. (Bekker et al. 2008, 3127)

From the very beginning, the government also formulated the Integrated National Electrification Programme (Barnes et al. 2009, 5). A 1992 National Electrification conference resulted in the creation of a National Electrifi-cation Forum, with the "main achievement ... to combine technical and financial capabilities with political legitimacy and support, and [to form] an arena where stakeholders could negotiate the shape of an electrification programme, which would be both politically acceptable and practically implementable" (Bekker et al. 2008, 3128). Notably, the conference itself was held under the banner of the leading antiapartheid political party, Nel-son Mandela's ANC, in "a deliberate political move to place electrification on the policy agenda in the context of the transition" (Marquard 2006, 180).

While government interest and institutional capacity were clearly the key drivers of South Africa's success, the role of local accountability mechanisms also warrants consideration. Here the evidence comes largely from the recognition that while the local administrative apparatus "was highly fragmented and faced many challenges, a few of the better managed municipal distributors made a significant contribution to the overall electrification effort" (Bekker et al. 2008, 3135). According to Eberhard (2004, 37), rural electrification achieved such an important status in South Africa because "urgency of promoting social equity and extending improved infrastructural services to the majority forced Eskom and the large municipalities to respond to the challenge of electrification." Indeed, one of the major changes in South African electricity distribution at the end of apartheid rule was increased "political attention" because "distributors in the poorest areas have been unable to finance new connections and subsidized services to poor customers that are pillars of the electrification programme" (Eberhard 2004, 16). In other words, municipal distributors were now facing increased political pressure to provide basic electricity services to their people, while needing to tailor these services to context-specific and individual household needs (Barnes et al. 2009).

According to the government's own evaluation of the rural electrification effort in 2001, in most cases local authorities played an important role in the process, especially in the following areas: "connection prioritisation/ scheduling ... identifying local labour for use by the implementer ... communication of implementation status to communities ... representing the community on issues of technology choice (mainly the connection capacity and metering options)" (DME 2001, 9). Dinkelman (2011, 3084) confirms that "local political pressures and connections costs" were crucial for this form of community-level electrification. Despite various challenges, distributors found community participation important for their success:

In general, all distributors considered that community participation efforts they engaged in were very worthwhile, and clearly contributed to implementation success and customer satisfaction, although such processes could be tedious at times. (DME 2001, 13)

Hard data bear out the importance of local accountability. Gaunt (2005, 1310) provides data for the national electrification program that began in 1994. Of the total annual targets between 1994 and 1999, which were either met or clearly exceeded every year, the share of new non-Eskom—in practice, municipal—connections was at least 25 percent in every year and up to 33 percent in four of the six years. According to the 2001 government evaluation cited above (DME 2001, 28), key factors in explaining the quality of cooperation between the electrification agency, whether Eskom or a municipal corporation, and community representatives were the capabilities and expertise of the representatives. This reflects in particular on Eskom's approach to understand electrification as a development strategy (Bekker et al. 2008, 3127). Local accountability is integral to this approach for providing both electricity services and rural infrastructure development more broadly.

In South Africa, new technology did not play an important role in successful eradication of energy poverty. Gaunt (2005, 1312) correctly notes that "the electrification was characterised by innovative research and development, and a continuous and substantial change of technological standards," but none of these changes were genuinely new; rather, the institutional capacity of the government and Eskom allowed South Africa to apply existing solutions. Examples included using single-phase instead of three-phase distribution to reduce costs, using prepayment metering, and following strict standardization guidelines across localities.

In conclusion, South Africa's success can be summarized as a natural experiment in government interest. As apartheid ended, South Africa's new democratic government was ready to initiate ambitious rural electrification programs, capitalizing on impressive institutional capacity and, to a somewhat lesser extent, relying on local accountability mechanisms in the rural communities. Given the low cost of electricity in South Africa, rural electrification also turned out to be a solution to the problem of access to modern cooking fuels: electricity is now the most common fuel for cooking even among the rural poor. South Africa's rural electrification program, clearly an expensive investment and a major challenge, succeeded in eradicating a lot of energy poverty in a relatively short period of time.

Ghana: Democratization and Energy Policy Reform

Similar to South Africa, the story of energy poverty in Ghana can be divided into two parts. The first phase of Ghana's struggle with energy poverty starts with its independence in 1957 and lasts until the early 1990s. Among many other African countries, Ghana suffered from a tumultuous series of coups and countercoups. Between the toppling of the first prime minister, Kwame Nkrumah, in 1966 and the arrival of Jerry John Rawlings to power in 1981, eight different leaders—mostly military—replaced each other and had a tenuous hold on power. Solving energy poverty was low on their list of policy priorities. Then in the early 1990s, Rawlings began a slow and imperfect transition to democratic institutions (Gyimah-Boadi 1994). Ghana's political transformation accelerated in 2000 when John Kufuor peacefully replaced Rawlings after relatively free national elections. Economic growth then picked up rapidly. According to the World Bank, income per capita grew from US$263 in 2000 to over US$1,500 per capita in 2016. In parallel, the democratic transition coincided with profound reforms of the energy market that achieved considerable success.

We classify Ghana as a partial success with respect to electricity and a failure with respect to cooking. The reforms undertaken in the late 1980s certainly improved Ghana's lot. Electricity was mostly generated from dams located on the Volta River, and these mostly benefited the population of the nearby capital city, Accra. The rest of the country was still lagging. At the time, Ghana hosted a population of 14 million people (28 million in 2016), two-thirds of whom lived in rural areas (about half now). Overall, according to Opam (1995, 6), only 35 percent of the population had access to electricity at the time. Other sources suggest even lower rates (Ghana Ministry of Energy 2010).

Two shocks in the late 1980s proved to be game changers. First, a series of droughts exposed the vulnerability of a power sector almost exclusively based on hydraulic sources. In 1983, power generation collapsed to about 30 percent of its level from three years earlier because there was simply not enough water in the dams (Edjekumhene, Amadu, and Brew-Hammond 2001, 7). Second, the democratic transition shifted political power from a small ruling elite to a broader population (Gyimah-Boadi 1994; Green 1995). The result of these two shocks was strong pressure from the population to reform the energy sector. In 1989, the government announced the National Electrification Scheme (NES), which sought to provide electricity to the entire population by 2020 (Edjekumhene, Amadu, and Brew-Hammond, 2001, 9). The pace of reforms accelerated in 1994 with a series of additional reforms that attempted to make this ambitious objective more realistic. Since the implementation of the NES, Ghana has experienced a radical increase in the pace of electrification. The aggregate electrification rate increased from about 35 percent in 1990 to 45 percent in 2000, and to about 71 percent in 2014, a remarkable change (Ghana Statistical Service 2014, 92). Urban electrification reached about 89 percent, while rural electrification remained below 50 percent (Kemausuor et al. 2011).

The deployment of modern cooking fuels stands in stark contrast to the relatively successful electrification effort. Here, Ghana has made only limited progress. The vast majority of the population still relies on solid fuels, such as wood and charcoal (UNDP 2004), and LPG remains scarce. Although data are scarce, it is believed that less than 10 percent of Ghana's population uses LPG for cooking (UNDP 2004; Afrane and Ntiamoah 2012). This is not because the government ignored the potential of LPG, however; the country launched the National LPG Programme in the early 1990s to promote the use of LPG, but with little success. The exception is the urban population around Accra, where about one-fifth of the population uses LPG (UNDP 2004, 4). Still, the rest of the country continues to rely on solid fuels (mostly wood fuel), suggesting that the timid efforts of the government failed to reduce the problem of cooking fuels.

Conventional explanations come up short when explaining the Ghanaian case. The country always knew it had plenty of hydraulic resources, and it mostly made good use of them. The issue, for a long time, was the lack of government interest in expanding access to these precious resources to the rest of the country. Initially, the government limited its efforts to provide electricity to power its industrial sector with little regard for the broader population. The topology of the country as a whole does not raise any major obstacle to the provision of energy in rural areas. The country is fairly flat, and the rural population represents only about 10 million individuals. Wealth on its own does not help us explain the Ghanaian case either. Admittedly, Ghana has recently experienced a period of sustained growth, and its annual income has reached about US$1,500 per capita. Yet the government undertook its successful rural electrification programs before this period, at a time when income was not growing. In fact, per capita output stagnated for the entire period of the initial reforms; lack of income growth did not seem to hamper the successful expansion of electricity to rural households. Therefore, it seems unlikely that income was the main constraint on efforts to alleviate energy poverty.

Let us now examine whether our theory sheds more light on this case. Government interest in energy poverty remained low between independence and the late 1980s. More precisely, the government's attention to the power sector was strongly related to its industrial interests. Basic electricity generation was provided by the Volta River Authority, founded in 1961 (other utilities were created later). Under its management, two dams were built on the Volta River: the Akosombo (1965) and the Kpong (1982) dams (Kapika and Eberhard 2013, 195). The Akosombo dam is particularly noticeable, since it created one of the world's largest artificial lakes and displaced about 75,000 people.[1] The construction of this dam responded to industrial needs, in particular that of VALCO, an aluminum company. In its early years, VALCO consumed about 60 percent of the electricity generated by the Volta River Authority (Kapika and Eberhard 2013, 203).

What is particularly striking is that, drought years aside, electricity supply was generally sufficient to respond to demand. In fact, Ghana has exported a significant amount of electricity to Benin and Togo since the 1970s and to Côte d'Ivoire since the 1980s (Edjekumhene, Amadu, and Brew-Hammond 2001, 4). This was lucrative, since the export tariff was five times larger than what utilities could charge local customers. Thus, the main problem was not wealth or lack of resources; rather, the successive autocratic governments simply did not care. Power was there; what was lacking was an incentive to provide it to Ghana's most remote population.

The situation changed dramatically during the 1990s. The government, mindful of the transformation and the stabilization of the country, finally sought to improve the power sector for household consumers. It undertook a series of reforms that tried to address the sector's main weaknesses. The era of reform that started in 1989 and accelerated in the late 1990s significantly improved access to electricity. The NES, started in 1989, was at the core of these reforms, and it was an ambitious project. Its main aim was to provide electricity to all communities with more than five hundred inhabitants by 2020, a tall order for a country with an electrification rate that was believed to be between 25 and 35 percent depending on the source (Kemausuor et al. 2011; Noxie Consult 2010).

These reforms were needed. Until that time, electrification was constrained by chronic lack of capacity, partly generated by politicized fees, a widespread problem in Ghana. As Edjekumhene, Amadu, and Brew-Hammond (2001, 2) observed, "For many years, poor financial performance and low productivity characterized many of Ghana's [state-owned enter-prises]." The lack of resources increased the reliance of Ghana on external funders to keep its NES efforts afloat, and even these funders soon started to lose patience. For instance, the World Bank requested that Ghana "depoliti-cise the tariff-making process and adopt a formula for adjustments" in exchange for its financial support (Kapika and Eberhard 2013, 208). In addi-tion, the NES failed to reach smaller communities. Until then, the NES, and the electricity market in general, was very much a top-down operation. The plan was to expand electricity to district capitals first and then, ideally, to spread farther based on commercial perspectives. This approach, however, penalized remote and sparsely populated communities.

To respond to these challenges, institutions were modified in two ways. Regulators attempted to solve the power sector problems by implementing reforms that improved capacity and adjusted the incentives of stakeholders by increasing local accountability. First, in 1997, the government created the Public Utilities Regulatory Commission (PURC; University of Ghana 2005, 25). One of PURC's main tasks is to monitor the fees imposed by utilities. As a nominally independent institution, PURC was less exposed to the kind of political capture that was common earlier. As we saw, this was problematic because fees did not reflect the costs faced by utilities, a clear source of inefficiency and therefore a threat to capacity.

Besides fee monitoring, PURC's responsibilities also include collecting and addressing customers' complaints. The commission itself underlines this role: "[PURC] welcomes views from the public, as well as consumers' comments and complaints about their utility service. Such comments and complaints do not only help the Commission to protect consumer inter-ests, but they are also an invaluable means of assisting the Commission to monitor the efficiency of the utility companies."[2] PURC therefore serves as an institutional platform that enables local monitoring and channels information to key regulators. This proved useful to consumers when, for instance, fees imposed by utilities increased by 60 to 70 percent from one day to another in 2016. After many complaints, PURC requested utilities to suspend the new billing fees.[3]

Second, the NES was augmented with the Self Help Electrification Pro-gramme (SHEP) in 2001 (Ghana Ministry of Energy 2010), which targets areas that are sparsely populated but not too far from the grid (Bawakyil-lenuo 2009). The rationale was to let local communities organize requests to connect their houses to the grid. These communities were requested to provide funds to buy the poles, and they had to ensure that households were ready to be connected. These jobs were often completed by local businesses that local leaders could more easily monitor. Once local communities had found enough money to cover 30 percent of the costs of electrification, the government would provide the rest (Elojärvi et al., 2012, 574). The money increases local stakeholders' incentives to complete the project successfully and helps to ensure its proper funding. In the case of SHEP, local accountability serves multiple purposes: it transfers information to electricity regulators, facilitates the mobilization of citizens, and monitors the efforts of private and public providers.

These reforms proved successful. Although their effects took time to materialize, electrification rates increased substantially. Furthermore, the sector as a whole became healthier. The Volta River Authority increased its revenues and managed to make a profit for the first time in 2009 (Kapika and Eberhard 2013, table 6.4). Nonetheless, it is important to put these successes in context. Accra and the rest of the country still suffer from occasional blackouts, and the country is still highly reliant on hydropower.[4] Given that it seems unlikely that droughts will be rarer in the future, Ghana is still vulnerable to shocks. These issues are related to underinvestments in infrastructure and additional power sources, both of which are political failures. And as we noted, rural electrification rates remain far from optimal. Nonetheless, these issues are not surprising considering how quickly demand has expanded over the past thirty years. To some degree, growing pains seem unavoidable. Still, these problems underscore the importance of the factors that we have highlighted; in particular, improving institutional capacity may be the key to solve Ghana's energy issues.

What about modern cooking fuels? The failure of modern cooking fuels to reduce reliance on wood fuel and charcoal can mainly be explained by lack of government interest. The government halfheartedly started a number of projects to change the situation. We mentioned the National LPG Programme that was kick-started in 1990. Yet the effect of this program materialized almost entirely in Accra. In fact, UNDP (2004, 4) noted that there was only a single filling station in the Upper East and Upper West (the rural areas farthest away from Accra and the coast), revealing a lack of interest. The government later attempted to incentivize LPG distributors to go beyond the greater Accra region by providing subsidies through the

Unified Petroleum Price Fund. This had little effect. Overall, then, LPG grew reasonably in urban areas but remained beyond the reach of the poor (only about 2 percent of households in the lowest income quintile use LPG) and the rural areas. As Kemausuor et al. (2011, 5145) noted, "The LPG drive was successful, [but] it is observed that patronage was skewed in favour of urban dwellers."

In sum, the case of Ghana is instructive for a number of reasons. Although Ghana's economy is doing well at the time of this writing, its success with respect to electrification predates its output growth. Lack of GDP growth, while certainly a handicap, is not an unsurmountable obstacle. Instead, the Ghanaian case clarifies how important it is for governments to show clear interest and to design policies that solve the twin issues of institutional capacity and local accountability. If the efforts to promote LGP and other modern cooking fuels remained lackluster, the same cannot be said about electrification. The government both mobilized local inhabitants to drive the rural electrification effort and improved the resource base of its utilities.

Success in the End Game: Brazil and Chile

Latin America's energy poverty problem is different from those in Asia and Africa. Thanks to high levels of development and urbanization, Latin America mostly left energy poverty behind a long time ago. However, rural areas have seen only limited progress. The case of Chile in particular shows how democratization can strengthen government interest in universal service delivery and, in combination with high institutional capacity and local accountability, contributes to success in the end game of energy poverty eradication.

Brazil: Electrification as a Constitutional Right

Brazil is clearly a success story when it comes to fighting energy poverty. When the Diretas Já movement toppled Brazil's military junta and the country started its transition toward democratization in 1985, electrification rates were admittedly already high. In 1990, about 85 percent of households had access to electricity, but the difference between urban and rural populations was dramatic.

While only two out of every hundred Brazilians in the country's cities lacked electricity access in 1990, only about every second Brazilian in the countryside had access to power from the national grid, so, the major electrification challenge for any government was one of rural electrification. Indeed, according to latest data from the *World Energy Outlook* (IEA 2017), overall electrification rates were rising to almost 95 percent in 2000 and full electrification in 2016, driven by an unprecedented growth in rural electrification from 55 percent in 1990 to 97 percent in 2016. This success is not limited to electrification alone; it also entails access to modern cooking fuels. The use of LPG as primary cooking fuel is common in Brazil (Lucon, Coelho, and Goldemberg 2004, 82) and has been driving down the use of fuelwood as the primary cooking input from 74 percent in the 1970s to as little as 17 percent in 2000 (Geller et al. 2004).

Conventional explanations like geography do not completely explain Brazil's electrification levels. Being the fifth largest country in the world, Brazil has varied geographic and climatic conditions that have contributed to clear differences in population density. Urban centers like Rio de Janeiro and São Paulo have some of the highest population densities in the country, whereas Mato Grosso and Amazonas have the lowest population density numbers. Yet this has not stopped Brazil from making huge strides in ensuring that all its citizens have access to electricity, as the country achieved 99 percent electrification by the 2010 census. Thus, although high levels of urbanization help with providing electricity access in general, geography alone fails to explain changes in electrification levels in the country.

Relatively high income levels, especially among the urban rich, were helpful in facilitating transfer payments to finance rural electrification and generous LPG subsidies, but income alone cannot explain the timing of rural electrification well. Only when the Brazilian government strongly committed to actively push rural electrification programs in the early 2000s did electrification rates in rural locations rise. Brazil is rich in hydroelectric potential and oil, which is why the abundance of natural resources may be a potential explanation for its efforts to fight energy poverty, and it did help with achieving fairly high rates of access to electricity and modern cooking fuels fairly early on in the second half of the twentieth century. Yet resources alone are insufficient to explain the country's electrification success, as the primary challenge was to reach millions of households in remote rural areas.

In contrast, we argue that electrification levels in Brazil are rooted in strong government interest and unprecedented political commitment to lift its population out of energy poverty. Being the seventh largest economy in the world in terms of nominal GDP according to World Development Indicators and aspiring to modernity and the notion of becoming a world leader, the constitution of the República Federative do Brasil from 1988 assigns to the government the "full responsibility for the distribution of electricity as an essential public service" (Niez 2010, 19), This is the strongest stance a government can take to reassure its people of the commitment to reduce energy poverty by prioritizing energy access and ensuring its affordability (Jannuzzi and Goldemberg 2014). Setting Brazil apart from many other countries, this constitutional promise writes into law that "electricity access is recognized as a citizen right, which is quite unique a feature of the Brazilian approach to electrification" (Gómez and Silveira 2010, 6254). While the Brazilian approach to energy poverty is highly centralized at the federal level, strong government interest and, as we shall see below, high levels of institutional capacity were critical to this achievement. Supporting our theoretical claim further, these factors were so influential that they overcame problems of energy poverty in a country without the ideal geography because of its sheer size and remote habitations, especially in the Amazon and the Northeast. Importantly, our argument focuses on government interest, which is typically heightened by democratic institutions but can, of course, be present during authoritarian regimes. Even under the rule of the military junta in Brazil, President Emílio Médici in 1971, for example, presented the First National Development Plan that promoted economic development in Brazil's Northeastern Region and Amazonas. Similarly, President Ernesto Geisel, in an attempt to counteract slowing economic growth in the aftermath of the 1973 oil crisis and after exceptional growth rates of 10 percent in the early 1970s, known as the Brazilian miracle (*milagre econômico brasiliero*), invested heavily in the country's infrastructure, including in energy generation, transmission, and distribution systems. This effort, motivated by the leaders' interest to stabilize their authoritarian rule, helped to electrify the country.

Brazil's energy situation is rather favorable. Having almost quadrupled its electricity generating capacity from hydroelectric sources from 16 gigawatts in the 1970s to more than 60 gigawatts in the 2000s, Brazil generates 90 percent of its electricity from hydropower, which also covers almost half of the country's total energy use (Geller et al. 2004). Such high dependence did not always prove a blessing. In 2001, Brazil saw one of the most severe droughts the country had ever experienced, which required the government to take drastic action.[5] Energy consumption had to be cut by 20 percent. In order to achieve this goal, the Brazilian government offered cash incentives to those willing to save energy beyond this target, and penalized households not complying with these conservation targets to be disconnected from the grid for six days. Rationing measures, which also included dimming street lights at night and government employees working only six hours a day, were in place until May 2002 and are estimated to have incurred costs of about 3 percent of total GDP.[6] While the heat wave clearly triggered Brazil's energy crisis, the more fundamental problem was a lack of investment across the entire supply chain of electricity services. After Brazil's electricity sector was deregulated in the 1990s, electricity distribution in remote rural areas became difficult as private investors did not see profitable business opportunities (Niez 2010; Geller et al. 2004). Having stalled privatization in generation after the 2001 power shortage, electricity generation is now almost exclusively in the hands of state-controlled companies, first and foremost Eletrobras, which alone holds about 40 percent of generating capacity.[7]

Although the large and densely populated urban centers of Brazil, like São Paulo (13 million), Rio de Janeiro (6.5 million), and Salvador (3 million), benefit from centralized, large-scale electricity generation, providing electricity to those living in sparsely populated and remote areas has proven to be a major challenge to every Brazilian's constitutional right to electricity access. After more modest and largely unsuccessful attempts, such as the 1994 PRODEEM program to get renewable energy and water pumping to rural households or the Luz no Campo initiative to electrify 1 million rural households between 1999 and 2003, the Light for All (Luz para Todos, LPT) program was by far the most ambitious and far-reaching policy (Gómez and Silveira 2010; Niez 2010; van Els, de Souza Vianna, and Brasil Jr. 2012; Goldemberg, La Rovere, and Coelho 2004).[8] The program, which was initiated in 2003 under the auspices of the newly elected, and then incoming President Lula da Silva of the leftist Workers' Party, was established with a clear goal: to electrify every home. This program blends in with many of the Lula government's other social welfare programs, offering conditional cash transfers to the poorest families in Brazil. By 2010, according to the program's target, 15 million Brazilians in the countryside should have been granted full access to reliable electricity, no matter where they live. Highlighting the importance of this policy, Gómez and Silveira (2010, 6255) note that "communities that are located along the river side, far from existing roads or in environmentally protected areas, cannot be provided with electricity from the main grid," while van Els, de Souza Vianna, and Brasil Jr. (2012) argue that without government interest, universal rural electrification cannot be achieved in the country. Continued support for rural electrification and investments into maintenance and updating of the national grid were channeled through the 2007 Growth Acceleration Program (Programa de Aceleração do Crescimento, PAC), with more than 50 percent of the 500 billion reais budget (2007–2010) being spent on energy. A follow-up program, PAC-2, resulted in another US$600 billion being invested in energy projects across the country until after 2014 to secure reliable and affordable electricity access, promote the use of renewables, and benefit from end-use efficiency improvements.[9]

Rural electrification, in particular for states in the Amazon basin, has always been challenging. Brazil's electricity system consists of two separate grids, the interconnected system, Sistema Interligado Nacional, which connects the urban hubs in the southeast, along the Atlantic coastline up to the country's north, and several isolated systems in the Amazonas region that account for only about 4 percent of installed generating capacity. Since expanding the national grid was the favored means of pushing electrification forward, making the 2 million connections needed to get all rural households connected required major investments.

Notwithstanding challenges with electrifying periurban areas and slums, connection costs in highly urbanized areas are fairly low, while connection costs per household in Brazil's remote areas were soaring to US$4,000 per capita (Niez 2010). The government therefore had to fund the largest share of the US$7 billion rural electrification package itself in order to provide electricity access to rural villagers at no cost. This was made possible because Brazil's fee schedule increases gradually with consumption, which essentially implements indirect transfer payments. Richer households in Brazil's urban economic and commercial centers finance grid expansion and rural electrification; they partly also partly pay for the poor people's electricity bills, as rate increases are cut at 8 percent to avoid disproportionate burdens for deprived households (Niez 2010). These tariff subsidies bode well with the notion of solidarity in the Brazilian constitution's universal right to electricity (Zerriffi 2008) and the government's focus on access and affordability as highlighted by Jannuzzi and Goldemberg (2014); however, it also requires a large and affluent enough customer base to generate sufficient revenue for electricity utilities. Brazil's large urban population, which increased from about 75 percent in 1990 to 86 percent in 2016 according to World Bank data,[10] always enjoyed reliable electricity access and hence played an important role in enabling the government to build capacity and facilitate rural electrification.

Brazil's previous experience with distributed generation and solar-based electrification through its 1994 PRODEEM program was not positive. Consistent with our theoretical claims, the failure of the PRODEEM program can be largely attributed to a lack of institutional capacity and implementation that was too centralized and overlooked household preferences in remote villages. The major problem, besides incorrectly installed photovoltaic (PV) systems and a failure to provide technical support and sufficient maintenance, was that the 6,000 installed systems were too advanced a technology at that time to push rural electrification; additionally, all parts of the PV units were imported, undermining aspects of local sustainability and community management (Valer et al. 2014; van Els, de Souza Vianna, and Brasil Jr. 2012) and also drove up costs (Obermaier et al, 2012; Goldemberg, Rovere, and Coelho 2004). Despite PRODEEM's failure, today, more than two decades later, off-grid electrification appears to be the way forward to provide the remaining 1 million people in the Brazilian rain forest with electricity, as conventional grid expansion may not be feasible (Gómez and Silveira 2010; Niez 2010; Coelho and Goldemberg 2013; Jannuzzi and Goldemberg 2014). Distributed power generation would not only allow rural villagers in remote parts of the country to also enjoy the benefits from electrification (Pereira, Marcos, and da Silva 2010), but also enable politicians to target socioeconomic and developmental efforts more directly to lift the country's most destitute population out of poverty (Slough, Urpelainen, and Yang 2015).

In line with van Els, de Souza Vianna, and Brasil Jr. (2012), the LPT was perceived not only as a rural electrification program but also as a policy to promote sustainable development and strengthen local accountability. Gómez and Silveira (2010, 6235, emphasis added) highlight that this policy was targeted primarily at

(i) municipalities with electricity coverage lower than 85%, (ii) rural electrification projects that benefit people affected by dams, (iii) rural electrification projects that focus on the productive use of electricity and promote *integrated local* development, (iv) rural electrification projects in public schools, health centers and water supply plants, (v) rural electrification projects designed to meet *rural settlements electricity needs*, and (vi) electrification projects for the development of family farming.

This focus on local needs and attempts to accommodate local communities' preferences is an important aspect of our theoretical argument and contributed largely to the acclaimed success of the initiative. Accountability was an integral part in the implementation of the Light for All program. For one, reflecting the centralized nature of the program, Congress and state legislators, according to a World Bank report (Correa et al. 2006) exercise considerable control over regulatory agencies across the board, including the Brazilian Electricity Regulatory Agency (ANEEL). This oversight includes public hearings, summoning of agency directors, and requests for explanation, which creates upward accountability of the implementing regulator to the legislators. Accountability matters also in another way. The operational structure of the LPT and electrification priorities are explained in a handbook that is available to local communities. In practice, Gómez and Silveira (2010, 6254) argue this means that "citizens who still lack electricity service [can] request the connection to the electricity company." These requests are then fed into the national electrification expansion plan, and concessionaires (the electricity companies) are required to report directly to the Ministry of Mines and Energy upon completion of grid extension.

With compulsory electrification targets, the design of the LPT program generates high levels of local accountability paired with national oversight. Local communities provide information about electricity demand to utilities, which, by governmental decree, have to be responsive. Concessionaries also have the responsibility "to conduct educational and awareness campaigns to inform all the inhabitants of rural communities in their respective concession area about the possibility to request connection" (Gómez and Silveira 2010, 6254). This ensures that the voices of the electricity-poor population in Brazil's countryside are heard and that their preferences and needs shape how rural electrification is governed. Notwithstanding LPT's overall effectiveness in terms of connecting rural households, however, a more recent study by Slough, Urpelainen, and Yang

(2015) based on microlevel data suggests that the largest benefits of Brazil's ambitious rural electrification program accrued to relatively wealthy but still nonelectrified municipalities, as opposed to their poorer counterparts. Given Brazil's rugged terrain and vast size, this is likely the result of rural electrification that is based on grid extension. Yet for continued rural electrification, local accountability will be playing an even bigger role because distributed power generation is critically dependent on community preferences, local experience, and effective information transmission (van Els, de Souza Vianna, and Brasil Jr. 2012; Gómez and Silveira 2010; De Oliveira and Laan 2010).

While LPT is set up to leverage local accountability for successful implementation, Brazil's rural electrification program is a largely centralized initiative. Initiated under President Lula's first term, the aspiration was to electrify all Brazilian homes, near and far, and to live up to the country's constitutional promise (Goldemberg, Rovere, and Coelho 2004). Firmly embedded in Brazil's 1988 Republican Constitution, government interest in electrification could not have been stronger. With various governmental energy programs in place, like ethanol, biofuel, and biomass promotion schemes (Martinelli and Filoso 2008; Goldemberg, Coelho, and Guardabassi 2008; Lora and Andrade 2009; Sperling 1987), the creation of ANEEL in 2000 endowed the federal government with sufficient institutional capacity to regulate and implement rural electrification effectively (Gómez and Silveira 2010). Its powers include the determination of future energy exploration, the implementation of policies to monitor contracts with private power companies, and setting the criteria for rates. This federal agency also has cooperative agreements with Brazilian state governments to ensure that its activities are carried out efficiently.

A similar success of the centralized "large-scale approach" (Coelho and Goldemberg 2013, 1088) can also be observed in how Brazil improved energy access to modern cooking fuels. Access to LPG in Brazil is unrivaled, with access rates of 98 percent for all households and 93 percent for rural ones (Lucon, Coelho, and Goldemberg 2004, 82). Again, these figures are spectacular given the country's size and difficulty to access even remote areas in the rain forest. Initially, cooking with gas benefited only urban households in Brazil's major cities because it required gas pipes. As early as 1937, Brazil's domestic use of LPG caught up considerably, admittedly more by accident than design.

When all Graf Zeppelin trips to South America were canceled due to the Hindenburg fire in 1937, this left behind cylinders of propane gas, which were distributed to 166 households, and Brazil's LPG "program" was born (Lucon, Coelho, and Goldemberg 2004; Coelho and Goldemberg 2013). After importing large amounts of LPG from Europe, the United States, and Argentina during World War II, Brazil started its own production of LPG in the mid-1950s. During Brazil's Second Republic, President Getlio Vargas promoted import substitution as a strategy of industrialization and sought to develop national champions to power the country's economic development, with Petrobras becoming a key player among them. Apart from monopolizing LPG production in the hands of state-owned Petrobras, the government also facilitated the creation of a national retail market (Jannuzzi and Goldemberg 2014, 259). A company now known as Ultragaz, controlling a quarter of Brazil's LPG distribution market in 2015, entered and expanded its distribution business, already reaching over 1 million customers in 1960.[11] The ease and convenience of using LPG for cooking created significant demand, and even poor households could afford it because of government subsidies (Jannuzzi and Sanga 2004). "Energy for cooking is a basic need and as a public good should be affordable and equitably distributed among all households, irrespective of their income and location" write Jannuzzi and Goldemberg (2014, 259). This offers great insight into where any Brazilian government's interest, over the past four decades, for subsidizing energy access to modern cooking fuels is coming from. Ultimately we saw a phenomenal rise in LPG customers, leading to 150 million Brazilians today using LPG on a daily basis, with related employment, social, and health benefits (Pereira, Freitas, and da Silva 2010; Lucon, Coelho, and Goldemberg 2004). Coelho and Goldemberg (2013, 1091) show that by 2010, the use of firewood as primary cooking fuel, particularly in rural areas, was cut by two-thirds relative to the 1970s. Used mainly as a cooking fuel in the north and northeastern states, LPG is also used for heating water and houses in the richer South, multiplying the fuel's effects for the Brazilian people.

Two factors played a key role in enabling widespread access to modern cooking fuels in Brazil. One is that the Brazilian government was committed to investments in the infrastructure while subsidizing and adjusting LPG pricing so as to make it accessible to even the poorest. Notably, this has been true for both democratic governments of Vargas (1951–1954),

Juscelino Kubitschek (1956–1961), and João Goulard (1961–1964) during the Second Republic, as well as under authoritarian, military rule from 1964 to 1988, for essentially identical reasons: to promote economic growth by building strong national companies from available resources and to stabilize political rule and garner popular support. For half a century, from the 1950s through 2001, the federal government regulated and heavily subsidized LPG prices in Brazil, resulting in expenditures of over US$100 million per year in 2000 (Lucon, Coelho, and Goldemberg 2004, 86). As this practice stalled investments from new entrants into the market and drove up foreign imports due to bottlenecks in Brazil's own production, the country's trade balance suffered, ultimately requiring adjustments to pricing. Once the subsidy was removed in the early 2000s, prices for LPG rose by 20 percent and led the government to introduce a "gas assistance," Auxílio Gas (Law 10453), for families earning 50 percent of the minimum wage. This voucher-based subsidy, which became integrated into Lula's 2003 Bolsa Família social welfare program, continued the support for households with low incomes that would be most likely to revert to firewood and more traditional biomass for cooking. Although the use of traditional biomass slowly increased in times of market liberalization, starting in 2006 with the onset of extended economic growth, traditional biomass declined considerably again. This trend was also strengthened as average income grew and urbanization intensified. Critical to the widespread use of LPG in Brazil, amounting to a total expenditure of US$8.24 billion between 1973 and 2001 (Jannuzzi and Sanga 2004), were the government's continued subsidies over decades (De Oliveira and Laan 2010; Lucon, Coelho, and Goldemberg 2004).

The second feature of Brazil's LPG program is high institutional capacity in the form of an excellent LPG distribution infrastructure. LPG is an ideal cooking fuel in Brazil because it is readily available and rather easy to maintain and transport:

Bottles with LPG (13kg bottles) are distributed by private companies all over the country and sold with affordable prices (initially they were subsidized), *even in Amazonian remote villages*. This 13kg bottle is available in several specialized stores or distributed by trucks or boats (in Amazonia). LPG delivery infrastructure is highly developed in all regions, *including rural zones*. (Coelho and Goldemberg, 2013, 1091, emphasis added)

Guided by the Brazilian Constitution's aspiration to eradicate energy poverty in the country, even remote villages are served with LPG, which enabled rural communities in the Amazonas to replace "existing fuel wood stoves … with LPG stoves" (Coelho and Goldemberg 2013, 1091). This "automatic delivery system" (Lucon, Coelho, and Goldemberg 2004, 85) guaranteed the distribution of LPG fuels around the country and hence must be considered an important pillar in Brazil's effort to provide energy access. To set up, run, and maintain such an infrastructure requires high institutional capacity and administrative skill. While distribution has largely been in the hands of private enterprises since the 1960s, the distribution infrastructure was centrally built with the help of the federal government by protecting the market from foreign entrants and guaranteeing profit margins for retail and distribution companies (Jannuzzi and Goldemberg 2014). This governmental vision was complemented with energy infrastructure investments that have always been present in Brazilian history, be it as part of the Salte Plan under President Eurico Dutra in 1949 (Singer 1953), the First Development Plan in 1971 under Médici, or, more recently, the Growth Acceleration Programs under Presidents Lula and Dilma Rousseff.

Even challenges with the implementation of the Brazilian LPG program testify to institutional capacity. The success of the LPG subsidy program created a problem of diversion to uses "such as heating swimming pools and saunas. Although the practice was prohibited, there were many vehicles utilizing it in clandestine and potentially dangerous adaptations" (Lucon, Coelho, and Goldemberg 2004, 86). In 2002, the subsidies were replaced with a cash transfer system for the poorest households to reduce leakage of LPG subsidy into nonintended uses (Lucon, Coelho, and Goldemberg 2004, 86). According to three Brazilian policy experts at the federal senate, whom we interviewed in August 2014 in the country's capital, Brasilia, the system has been a success: poor households receive assistance that allows them to continue the use of a clean cooking fuel, with the associated environmental and public health benefits, while the replacement of a price subsidy with a cash transfer has stopped leakage.

We conclude the Brazilian case study with two main messages. First, we have seen that strong government interest over decades, present during both democratic and autocratic episodes in the country's history, coupled with effective institutional capacity, were successful in addressing energy poverty in one of the largest and most populous economies.

The aspirational goal in the country's constitution to provide energy to all Brazilians applies to both rural electrification and LPG distribution efforts. The second interesting aspect is that although important decisions were largely made at the federal government level, the rural electrification program was nonetheless designed to leverage local accountability. Needs and preferences of local communities are continuously incorporated into the national expansion plans by allowing rural dwellers to request grid connections from the electricity providers in their respective concession areas. Brazil hence combines the "large scale approach" (Coelho and Goldemberg 2013, 1088), which benefits from centralization through pooling resources and economies of scale, with local information. This approach can work when the institutional environment is right. Both the LPT and LPG programs could be implemented because the large and relatively wealthy urban population helped the government finance programs for universalizing energy access.

Chile: Success in Rural Electrification under Privatization

From the perspective of a contemporary observer, our decision to include Chile, a rich country and an OECD member since 2010, may be a surprise. Given Chile's economic success, it is not immediately obvious why a political economy analysis is necessary. However, as we shall see, this case offers valuable insights into the struggles that even rich countries face when tackling energy poverty. Historically, Chile's first energy policies go all the way back to the 1930s and the creation of the Corporation for the Development of Production (Mariía 1945). Its history has been tumultuous. The power sector struggled throughout the 1970s, and even the liberalization of the electricity sector by Augusto Pinochet in 1982 did little to solve the problem. While unbundling and breaking up utilities improved the efficiency of the power network in urban areas, the situation remained problematic in rural areas, where only 50 percent of rural households had electricity in the early 1990s (IEA 2009, 186).

In fact, Chile's success in rural electrification is quite recent. After years of poor performance, a new series of reforms that started in 1994 set the stage for a period of rapid growth. It is then that the government launched the Programa de Electrificación Rural (PER, National Program for Rural Electrification), which offers subsidies and material support to expand rural electrification to the country's most remote areas. The PER proved to be a tremendous success: by 2008, electrification rates in rural locations reached 95 percent, while the rate was 99 percent for the entire country (IEA 2009, 14).

Here it is useful to translate these rates into actual numbers of households. In 1990, about 1 million people living in rural areas were deprived of electricity. Less than twenty years later, that number had melted to about 100,000, a remarkable success story. We therefore can classify Chile as a partial failure until the late 1980s and a clear success since then. As we show, we can tie the PER and its success to the political changes that took place in the early 1990s, a period marked by a transition from Augusto Pinochet's harsh dictatorship to Patricio Aylwin Azócar's democracy.

The Chilean case cannot be explained with traditional arguments. To be sure, Chile has access to oil, coal, and gas, and it also has considerable hydraulic potential. However, access to these resources is not directly relevant to the challenge of last-mile connections for remote rural communities. Furthermore, while it is true that Chile's geographic situation makes some regions hard to reach, the size of its rural population—2 million people, or less than 20 percent of the population—is relatively small.

Most important, Chile's path cannot be attributed to its wealth. The country had resources and indeed spent some of them on providing electricity to some parts of its population, but the ultimate failure of early efforts to achieve universal electrification can be attributed to poor policies. Before Pinochet, the power sector was highly inefficient. In the words of Pollitt (2004, 3), "Chile's electricity utilities were in a mess. Inflation, high fuel prices and price controls on final prices had led to large losses and a lack of investment under public ownership." Later, under Pinochet, the problem was a lack of interest in the plight of the rural areas.

Chile's recent history of energy access can be split into two parts. The first era covers Augusto Pinochet's time in power. Pinochet replaced Salvador Allende in 1973 and stayed at the helm of the country until 1990. This time was marked by his tight grip on power and an emphasis on privatization and free market. The power sector was no exception, though not much happened on that front initially. Things changed rapidly in 1982, when the General Law of Electric Service expanded Pinochet's agenda to the energy sector. The General Law foresaw the breakup and the privatization of the entire power sector. Endesa, the main state-owned utility, was split into fourteen firms and sold to private investors. The electricity market became more efficient: the generation, transmission, and distribution of electricity all improved (McAllistair and Waddle 2007, 199). This would prove to be very useful in making rural electrification programs a success, but in the short run, these reforms did little to help meet rural demands. The reason is straightforward: private utilities could barely make a profit in remote areas (McAllistair and Waddle 2007, 207). These regions were generally poor; in fact, in 1980, about 34 percent of the rural population lived below the poverty line (against 16 percent in urban areas) (Psacharopoulos et al. 1995, 248). Pinochet's very emphasis on privatization meant that rural electrification would remain a problem. Utilities had few reasons to cater to this segment of the population.

The effect of Pinochet's 1982 reforms on energy poverty was therefore limited. Urban electrification rapidly reached high levels, but these developments masked systematic shortcomings in rural areas. By the time Pinochet was removed from power in 1990, only half of the rural population had access to electricity. Clearly the 1982 reforms had done little to solve the issue outside the country's major cities. The situation radically changed with the advent of a new democratic regime. Under President Patricio Aylwin and later under Eduardo Frei, the government started to pay closer attention to the energy needs of Chile's rural areas.

Frei's tenure in particular was instrumental. In 1994, he launched the National Program for Rural Electrification (PER), which was to be managed by the National Energy Commission, an institution that was staffed by both specialists and key stakeholders (Forcano 2003, 40). National priorities had shifted: the government was now serious about tackling energy access in rural locations. Frei aimed for rural electrification rates to reach 75 percent by 2000. Ricardo Lagos, his successor, aimed even higher, with a target of 90 percent. By publicly revealing these ambitious goals, these policymakers engaged their reputation if the PER and related policies were to fail. It is clear that government interest in rural electrification was now much higher than under Pinochet. Fortunately for Frei and Lagos, the PER turned out to be extremely successful. At the beginning of the 1990s, rural electrification stood at about 50 percent. By 1999, it had increased to 76 percent, slightly above Frei's target of 75 percent (Jadresic 2000, 76). And Chile did not slow down, as Lagos's own ambitious 90 percent aim was also met by 2008, when the rural electrification rate actually reached 95 percent (IEA 2009, 195).

Still, government interest is a necessary but not sufficient condition for rural electrification. As it turns out, both institutional capacity and accountability played a key role in the success of the PER. To understand this, we briefly examine the main features of the PER. First, the program was managed and overseen by the National Energy Commission, which included both specialists, such as engineers, and politicians. This design ensured that it had political capital and technical expertise. Since one of the commission's tasks was to set fees for rural customers, the dual presence of experts and politicians ensured a reasonable price, which put the program on a solid financial footing. Reliable resources from the government and from international donors such as the Inter-American Development Bank ensured that enough money would underpin electrification programs (McAllistair and Waddle 2007, 210).

Perhaps more important, all stakeholders had to contribute financially to the electrification effort. Jadresic (2000, 78) argued that "to ensure sustainability, all participants–the state, the electricity companies, and the users–would contribute to the funding of investment projects. The state's participation was needed because rural electrification projects usually are unprofitable for electric utilities, as a result of low electricity consumption, the distance from distribution centers, and the dispersion of dwellings. But state subsidies would be allocated only to projects with a positive social return." Beyond the state, customers were also required to pay a share of the electrification cost (IEA 2009, 200).

Still, to avoid discouraging poorer clients, distribution companies could lend the amount needed to cover the customers' share and to be repaid later. The rest of the costs would cover the expenses of the distribution company. Overall, this system ensured that the electrification effort would remain well funded. Capacity had proved to have been a major problem until the late 1970s, leading Pollitt (2004, 3) to claim that "inflation, high fuel prices and price controls on final prices had led to large losses and a lack of investment under public ownership." Clearly the design of the PER overcame this problem.

Local accountability also contributed to the success of the PER. The program was very much a bottom-up one. Jadresic (2000, 78) underscores this: "To ensure appropriate technology choices, promote local commitment and sustainability, and fit the new decentralized structure, the [PER] program designers decided that the regional governments would identify

needs, choose the solutions, and participate in the decisions on the allocation of central funds." Local accountability operated in different ways. First,
communities had to organize with distribution companies to come up with
an electrification plan (Forcano 2003, 40). Second, local authorities were
closely involved in all the steps of the program, including implementation.
Since mayors are locally elected, this offers communities with a powerful
tool to ensure due diligence by local authorities (McAllistair and Waddle
2007, 211). Finally, local communities were also financially involved in the
process. McAllistair and Waddle (2007, 211) note, "Given that the [govern-
mental] subsidy [to obtain electrification] can only be authorized once for
each family, community members have a broad incentive to make sure the
most effective solution is selected."

In other words, local households and both municipal and regional
authorities were all closely involved in the deployment of appropriate
solutions. If needed, local communities could replace utilities, distribution
companies, and (presumably) their local representatives. In this context,
Pinochet's earlier reforms proved useful, for if a distribution company was
found to be lacking, others could jump in to fill the gap. Such possibilities
did not—and still do not—exist in countries such as Senegal with their
monopolistic market structures.

Whereas Chile has solved its rural electrification problem, it is lagging
with respect to cooking fuels. The Chilean population relies on two main
sources of fuel: a minority uses a local variant of LPG (LPG "aire"; IEA 2012,
19), but the majority relies on biomass in general and firewood in particular
(IEA 2009, 188). Exact numbers are difficult to obtain because Chileans use
wood fuel for both cooking and heating. Still, a consequence of this reliance
on wood fuel is a high deforestation rate, especially around urban centers
(IEA 2009, 184). In recent years, Chile has tried to reduce this dependence
by investing in natural gas; this has proven to be difficult, however, because
it has to import the gas from its neighbors, especially Argentina, which are
reluctant to allow the practice. LPG is problematic as well because it is too
expensive for customers. In both cases, progress to transit away from solid
fuels toward modern energy sources has proved to be slow.

Some observers suggest that cooking could be fueled with electricity
from hydroresources, but that would imply significant investments in additional hydropower.[12] Meanwhile, the government signaled its intent to support renewable energy sources for cooking, but this approach has yet to

yield significant results (IEA 2009, 204). At any rate, the Chilean experience underscores the importance of relentless government interest to provide energy access. The lack of interest from successive governments ensured that the country would continue to rely on biomass.

Some may contend that Chile's success in reducing energy poverty is not a surprise given its relative wealth. Others may point to the privatization policies that Pinochet followed. These explanations, however, are incomplete. The institutional reforms of the mid-1980s did ensure that generation, transmission, and distribution companies became more efficient. But until the mid-1980s, rural areas were systematically disadvantaged; private utilities and distribution companies had few incentives to cater to a rural population that was poor and hard to reach. The real change came under the new democratic governments of the 1990s, which provided reliable resources and constant political support for rural electrification. It is then that electricity access was gradually expanded across the entire country. The design of these reforms was crucial. It emphasized good governance and concerted planning between local customers and regional authorities. The results swiftly followed, with booming electrification rates. However, and not unlike other cases presented in this chapter, the progress with respect to cooking was slower. Cooking clearly was less of a political priority; it was also less of a problem given people's easy access to wood fuel and relatively efficient stoves that reduced the negative side effects of the use of biomass.

Assessing the Cases: Sources of Success

While there is no question that the three fundamentals of geography, energy resources, and wealth are important, their ability to explain variation in energy poverty is limited. In all of the success cases described in this chapter, we have seen clear evidence of government interest. Especially in rural electrification, the primary driver of success has been sustained government interest over time, be it for economic or political reasons—or both. Even in the wealthier and more industrialized Latin American countries, the last mile to universal service delivery was possible only after profound political changes. Vietnam, a very poor country ravaged by two decades of total warfare and with limited access to domestic energy resources, was able to move fast because of sustained government commitment.

If government interest is the cornerstone of success, then institutional capacity and local accountability mechanisms must come next. In isolation, government interest has not allowed governments to achieve their policy goals. In many of the cases, we see breakthroughs in the eradication of energy poverty with increases in institutional capacity. South Africa and Vietnam, for example, were able to realize their ambitious plans only when their power sector structures made the shift from being vertically integrated monopolies designed to provide for a small political-economic elite to leaders in universal service delivery. In Brazil, the provision of LPG for clean cooking has required decades of institutional policy and infrastructural development. Similarly, in countries showing strong government interest and institutional capacity, the evidence for the relevance of local accountability is clear: communities capable of participating in the planning of rural electrification and the provision of modern cooking fuels improved faster than communities with limited local accountability. Even the authoritarian regimes of China and Vietnam were ultimately dependent on local contributions for their success in rural electrification; in China, the same also holds for clean cooking fuels.

From a research design perspective, looking at cases of success is clearly not sufficient. At least in principle, it is possible that we find high levels of government interest, institutional capacity, and local accountability also in cases of failure. In the next chapter, we therefore summarize lessons from case studies of partial and total failure in the eradication of energy poverty.

6 Country Case Studies: When Success Eludes

In the previous chapter, we summarized our case selection strategy and presented cases of success in eradicating energy poverty. However, successful cases are obviously only half of the story. To test our hypotheses systematically, we now present results from case studies of failure. In Asia, we examine India's smaller neighbor, Bangladesh, a country with a history of political instability and, until recently, slow economic development. We also examine the case of Indonesia, a much wealthier economy endowed with abundant energy resources. We then travel to Africa, where we use the Kenyan and Nigerian cases to shed light on why energy poverty eradication has been so difficult for African governments and societies.

In the cases considered so far, we have seen more progress in rural electrification than in the provision of modern cooking fuels. To understand this component of our argument better, we conclude this chapter with the case of Senegal. This West African country has not performed well in rural electrification, but it has achieved exceptional success in the provision of LPG to both rural and urban households. This success stems from its major deforestation crisis that began in the 1970s. People relied on wood fuel and charcoal to cook, exposing the country to desertification in the Sahel. The government's robust response was to encourage people to adopt LPG. The Senegalese case thus strengthens our theory by highlighting the effectiveness of a motivated government in the provision of modern cooking fuels.

Before we begin the case studies, we must say a few words about the concept of failure. Measuring failure is difficult not only because the concept is always relative, but also because failure is often not tangible—easy to see—in the way success is. Success can be observed as rapid changes for the better, whereas failure manifests itself in the absence of clear changes.

To overcome this problem, we consider a case a failure if energy access (1) falls clearly below other comparable countries in the region, (2) is not rising in line with regional trends, or (3) should be much higher based on geography, resources, and wealth. In cases of partial failure, however, we may see some isolated successes. For example, the case of Bangladesh shows that rural electrification can succeed under some institutional arrangements while failing under others. For this purpose, the three conventional explanations—geography, resources, and wealth—again offer a useful benchmark.

Challenges in Asia: Bangladesh and Indonesia

Neither Bangladesh nor Indonesia can be considered a success story, especially in contrast to Vietnam or China. Although Bangladesh has achieved some success with rural electrification cooperatives and Indonesia's overall rural electrification rate has increased despite the inadequacies of its government policy, progress in rural electrification and the provision of clean cooking fuels has been disappointing,

Bangladesh: The Importance of Institutional Design
The case of Bangladesh is complex. On the one hand, energy poverty remains a burden on the country's economy, especially in the country's rural parts (Barnes, Khandker, and Samad 2011). The national electrification rate hovers around 75 percent and remains below 67 percent in rural areas, compared to rates of 89 percent and 81 percent as the regional average for developing Asia (IEA 2017). These numbers may even be overoptimistic; a study by Asaduzzaman, Barnes, and Khandker (2011, xxii) found that only half of the population of grid-electrified villages are actually connected. Myanmar aside, Bangladesh clearly compares unfavorably to its South and Southeast Asian neighbors. India (74 percent), Laos (85 percent), and Thailand (100 percent) all have rural electrification rates that are considerably higher. The situation is even worse when we consider cooking, because the country's rural population relies almost entirely on biomass (Barnes, Khandker, and Samad 2011; Mobarak et al. 2012). Even when people use nonbiomass fuels, they typically use kerosene (Asaduzzaman, Barnes, and Khandker 2011, 12) instead of LPG, which is virtually nonexistent in that country.

Yet Bangladesh's rural electrification policies over the past four decades are often hailed as an example that other developing countries ought to follow (Taniguchi and Kaneko 2009). For instance, a US Agency for International Development report argued that the rural electrification model of Bangladesh was a "best practice" and "is recommended ... to sustain or replicate in other countries of the region" (Gunaratne 2002, v).[1] On top of this, Bangladesh became a pioneer in off-grid deployment of renewable energy, in particular photovoltaic solar home systems (Hossain and Badr 2007; Barnes, Khandker, and Samad 2011). Today, millions of households have access to electricity through these new technologies.

How can we make sense of what looks at first sight like a contradiction? To begin, poverty certainly constrained policymakers. The country's GDP per capita was about US$117 around the time of independence in 1973, and even in 2016, that had increased to only about US$1,300, placing Bangladesh behind India and Pakistan and among the poorest countries worldwide. And yet the story of Bangladesh cannot be reduced to poverty. As we will see, Bangladesh implemented a policy that looks like a large-scale experiment: its government ran distinct electrification programs in parallel. One became relatively successful, though at a fairly small scale, while the others failed. The fact that one approach did succeed tells us that low GDP figures cannot, on their own, explain the failures that Bangladesh experienced. Neither can failures be explained by a lack of resources: Bangladesh has natural gas, although the deposits have long remained underexploited. And geography certainly cannot explain the outcome: exceptionally high population densities should make rural electrification in Bangladesh easy.

As we look for a political economy explanation, we start with electrification, which is both more complicated and interesting than cooking. To understand rural electrification in Bangladesh, it is useful to go back to the roots of current policies and institutions. Electrification has been a political priority since the country's independence in 1972, and policymakers have applied three distinct sets of policies to solve the problem.

The first was a traditional top-down approach. The government funded the construction of new plants, which were then run by the Bangladesh Power Development Board (BPDB, created in 1972), the Dhaka Electric Supply Authority (DESA; 1990), the Power Grid Company of Bangladesh (PGCB; 1996), or the Dhaka Electric Supply Company (DESCO; 1997). All of these institutions and companies turned out to be highly inefficient and unable to improve electrification rates. Their inefficiency revealed a lack of institutional capacity, which was caused by widespread corruption (Taniguchi and Kaneko, 2009). The absence of accountability mechanisms ensured that they remained chronically underfunded and poorly run. The arrest in 2007 of Sheikh Hasina, a former prime minister of Bangladesh, for extortion related to the construction of a power plant exemplifies the corroding effects of corruption.[2]

However, Bangladesh has also run some much more successful programs. The best known of these is the Rural Electrification Board (REB), founded in 1977. Unlike the BPDB and the DESCO, the REB has performed well within its limited means. As of 2010, almost 50,000 villages out of the 84,000 that had been targeted earlier were connected to the grid by REB (Rahman et al. 2013), and these numbers have continued to increase since then, though more slowly. Two crucial reasons for this success appear to be the ability to raise enough resources (high institutional capacity) and a systematic involvement of local communities (high local accountability).

A defining feature of the REB is its bottom-up approach to electrification. Local communities are organized in cooperatives called Palli Bidyut Samity (PBS), of which there are about seventy in the country. A PBS buys power from large suppliers and runs the distribution to local consumers, who own the PBS as a cooperative. The consumer-owners elect the governing body of their PBS. They select candidates among fellow members, which means that only local stakeholders are allowed to have an operational role (UNDP 2009, 25). To thwart political capture, politicians are barred from becoming board members (Taniguchi and Kaneko 2009).

At the top of this system, the REB's input is limited to the creation of new regulation and monitoring the PBSs. They also verify that the revenues are large enough to fund investments and maintenance, thereby guaranteeing high capacity (Yadoo and Cruickshank 2010). The REB provides technical assistance and financial support to the PBSs. The entire process is characterized by extreme transparency. For instance, the PBS of Thakurgaon (Rangpur Division, northeast) has its own website, listing all electricity rates for private consumers, commercial firms, and charities.[3] It also contains all of the forms required to become a new customer, information about current issues, and so forth. Finally, all officers are listed with their phone numbers and email addresses.

Because PBSs are well funded and consumers are closely involved in the decision-making process, this system has been particularly successful. Distribution losses are about 16 percent, half the national rate (Yadoo and Cruickshank 2010, 2943). Unsurprisingly given the role played by local communities, collection rates have also been high (up to 96 percent). Overall, the REB's success is often attributed to its independence from political meddling, its high efficiency, and its ability to engage with customers (Barnes 2011).

The third component of the government's approach to electricity access consists of off-grid solutions. Unlike large-scale grid expansion, these new technologies are often developed and deployed by enthusiastic private entrepreneurs, and Bangladesh illustrates the transformative effects that these technologies can have. In particular, solar home systems have proved to be a game changer. While the numbers keep changing, it is believed that by 2014, about 3 to 4 million households had adopted a solar home system, far above earlier targets of about 1 million (Palit and Chaurey 2011; Marro and Bertsch 2015) and far above the 45,000 units in operation around 2005 (Uddin and Taplin 2009).[4] In other words, about 15 to 16 million people have electricity through solar home systems, a staggering number.[5]

Overall, we therefore consider Bangladesh to be a partial failure. Its policies failed whenever they followed a rigid top-down script that suffered from inefficiencies and corruption. The failure of these policies is reflected in the large number of people who still lack basic access to electricity. However, we consider Bangladesh to be only a partial failure because the policies that did take into account the importance of local accountability and high capacity were successful for the most part. By looking at two kinds of policies in the same country over the same time period, we can see that the bottom-up, REB-backed policies were successful while the top-down programs failed.

The cooking situation in rural Bangladesh warrants only a brief digression. Little progress has been made to reduce the dependence of the rural population on firewood or other biomass products. LPG is virtually non-existent, and modern biogas and improved cookstoves are extremely rare. Regardless of capacity and accountability issues, this lack of progress mostly stems from an almost complete absence of government interest in the cooking problem. We found very little evidence of any large-scale program designed to change the situation, and the few that existed generally failed.

Given the lack of action on cooking, we dedicate the rest of this case study to understanding the intriguing variation in rural electrification. To begin with government interest, officials have, for the most part, been consistently supportive of rural electrification policy: government interest has generally—but not always—been high. Barnes and Foley (2004, 13) contend that the "government strongly supports the rural electrification program ever since the founding of Bangladesh."

The evidence supports this claim. Bangladesh's original 1972 constitution specifically spelled out the importance of rural electrification. Electrification is listed as one of the "fundamental principles of state policy," and Article 16 requests that "the State shall adopt effective measures to bring about a radical transformation in the rural areas through ... the provision of rural electrification."[6] To follow up on this requirement, one of the new authorities' first decision was to create the BPDB as soon as the country became independent in 1972. The BPDB's role was to plan the creation of a modern electricity infrastructure for the entire country. It soon appeared, however, that the BPDB was failing to make any significant progress in tackling electrification issues. For instance, electricity theft remained prevalent. Instead of keeping the status quo, authorities sought to solve the problem by adding new institutions, such as the DESA in 1990 (focused on Dhaka, Bangladesh's capital) and the PGCB and DESCO a few years later.

In parallel, in an effort to improve the situation, the government published the Rural Electrification Board Ordinance in 1977, which eventually led to the creation of the REB (Franda 1981). This ordinance had been requested by Ziaur (Zia) Rahman, who had become president earlier that year—the country's seventh in six years since independence. This piece of legislation was described by some as an exercise in state and democracy building (Bertocci 1982). As it turns out, the ordinance was quite powerful, underscoring the high level of interest from the government for electrification. For instance, paragraph 20 states that

any land required by the Board for carrying out its functions under this Ordinance shall be deemed to be needed for a public purpose and such land may be requisitioned or acquired for the Board by the Government or the Deputy Commissioner, as the case may be, in accordance with any law for the time being in force.[7]

And while the REB's board was mostly staffed with individuals appointed by the government, it remained somewhat independent from political interference. This means that the government was willing to delegate a fairly powerful major right (land confiscation) to a third party.

Government support for the REB has remained constant, though the temptation to capture it for political gain must often have been high. Indeed, despite its successes, the scale of REB operations has remained limited (Taniguchi and Kaneko 2009). According to a 2009 report, the total number of private consumers at the time was estimated at about 7 million people (Rural Electrification Board 2009, 13). Furthermore, it is difficult to assess the precise level of interest for certain time periods. For instance, after Zia's assassination in 1982, four presidents successively took power over a period of about six years. Undoubtedly, military leaders were busy scheming to stay in power or take it from their opponent. Nonetheless, it is remarkable that the basic structure of the REP did not suffer from political meddling during this period.

In the mid-2000s, observers started to worry about political interference in REB operations. Taniguchi and Kaneko (2009) document how the REB's very success seemed to have emboldened policymakers to interfere with its operations for personal gain. One result was a significant decline in the rate of new villages being electrified under the REB toward the end of the 2000s, going from a peak of 3,463 in 1999 to a paltry 364 in 2007 (Rural Electrification Board 2009, 15). Rahman et al. (2013) noted a similar slowdown.

This problem points to the critical role of institutional capacity and local accountability. The dangers of corruption become particularly salient under weak institutions and poor accountability structures. Again, it is useful to note the dual system of rural electrification in Bangladesh. The institutions characterized by strong top-down management have fared extremely poorly on both capacity and accountability. A 2004 study shows that the transmission and distribution losses of BPDB (28 percent), DESA (30 percent), and DESCO (33 percent) were very high (Alam et al., 2004, table 5).

In contrast, the REB is doing remarkably well, with transmission and distribution losses limited to 16 percent, and we contend that both capacity and accountability play a key role. In terms of capacity, the PBSs not only are less bureaucratic and benefit from their autonomy (Khandker, Barnes, and Samad 2009) but also have less corruption (Rahman et al. 2013). As Khandker, Barnes, and Samad (2009, 5) contend, the REB produces superior results because the usual distributional problems that plague other Bangladeshi distributors (theft, non-payment of bill by influential subscribers, illegal connection, overbilling, etc.) are almost nonexistent in the operation of REB. REB also has an almost perfect bill collection record—over 95 percent of the REB customers pay their bills. REB has improved further since 2000, and reported a 13 percent system loss during 2005–2006 (REB 2007), which compares quite favorably to the system loss in other South Asian countries. The success of REB is due mostly to its autonomy, minimal bureaucracy, strong culture of integrity, donor support and trust, and strong and independent leadership.

Accountability matters as well. As we saw, one of the key features of the PBSs is that their management is selected by their local customers from their own ranks. As long as social punishment mechanisms function, each PBS governing body has a strong incentive to perform well. Furthermore, politicians are not acceptable candidates, which precludes political capture of these resources. Rahman et al. (2013, 845) explain the importance of local accountability:

> Community participation has been an important factor contributing to the success of rural electrification. Every electricity user is a member of a rural cooperative and has the right to be involved in the decision-making and policy-making practices of the cooperative through their elected representatives, which are called directors. This membership practice gives the electricity users a feeling of ownership in the cooperative and encourages them to protect the assets from thieves and abuse.

Customers are directly involved in the management of their PBS, and this practice has two benefits. First, it encourages them to ensure that common problems that plague electrification elsewhere, such as theft, remain limited. The cooperative setup, supported by the REB, is self-enforcing because it is in the interest of each member to ensure that other members pay their dues and that nonmembers are prevented from stealing.

Second, customers have the power to remove ineffective operators from their PBS. As we said, only members can be elected to the board of directors. If they prove to be unable to complete their task, they can be replaced by the members themselves. This avoids the political tampering that plagues the management of many electrification agencies. Inefficiencies such as theft are easy to observe. Holding the local PBS's management accountable is therefore relatively easy, at least compared to having to remove a bureaucrat from the national government.

Do these variables help us understand the success of off-grid solutions such as solar home systems? It is not entirely surprising to see that new technologies became instrumental in Bangladesh. As we noted, the government's interest in electrification ranged from mild to high over the past three decades. The problems, such as corruption and low accountability, came from the implementation of particular policies and institutions. To the extent that these can be bypassed, entrepreneurs of all sorts began to see Bangladesh as a fertile ground to spread new technologies. As we saw, solar home systems became particularly appealing. One reason was that they are relatively easy to operate and affordable, making them ideal in Bangladesh.

Historically, the push for solar home system adoption in Bangladesh was greatly facilitated by Grameen Shakti, a local renewable energy firm. Grameen Shakti is part of the broader Grameen conglomerate founded by Muhammad Yunus, which also includes Grameen Bank, a microfinance bank. The model is simple (Amin and Langendoen 2012): Grameen Shakti draws on good relations with local communities to sell affordable solar home systems, and if the potential customers need a loan, they can apply for one from the Grameen Bank.

There are many reasons behind the success of the Grameen model. Its management has proved to be competent, and its leadership has skillfully enlisted supporters both at home and abroad. Additionally, Grameen's approach to capacity and accountability issues was innovative. Its financing structure ensured that the entire operation rested on a solid foundation. Grameen was never a charity; it always aspired to run a profitable business. The difference with other programs is that when customers faced problems, Grameen provided solutions, such as loans.

Similarly, many have claimed that the success of Grameen Shakti can be attributed to its strong culture of accountability (Sovacool 2012a). Whereas many big players in off-grid electrification keep their customers at arm's length, Grameen kept deep ties with local leaders and customers. Thus, information about events on the ground kept everybody honest.[8]

The important role of the private sector should not hide the importance of the government. One of the critical obstacles that decentralized technologies such as solar home systems face is financing. The Infrastructure Development Company Limited (IDCOL) was founded in 1997 by the government to provide financing to nongovernmental organizations and small entrepreneurs operating in the renewable energy business, including Grameen (Islam, Islam, and Rahman 2006). In some cases, IDCOL also subsidizes the technologies it promotes to increase adoption rates. Although

IDCOL is owned by the government, it is corporatized and operated independently (Uddin and Taplin 2009).

IDCOL is generally acknowledged as a key contributor to solar home system deployment in Bangladesh. It is difficult to assess the exact level of government interest in its success, however, because the institution was designed from the outset as an autonomous entity to collaborate with the private sector. Nevertheless, a number of signs suggest that the government was supportive of IDCOL's efforts; the organization benefited from a high level of capacity, and local accountability was quite high throughout.

First, the government provided the initial cash for IDCOL's operations in 1997 (Marro and Bertsch 2015, 2). In 2005, it added more capital to help it expand its operations. The government also reduced import tariffs on important components for solar home systems, which again suggests that it was willing to help IDCOL (NREL 2015, 6). High levels of capacity were important as well. On top of the government's own resources, IDCOL benefited from generous aid from the World Bank and other international development agencies (Samad et al. 2013). These sources ensured that cash constraints would be less binding. Equally important, IDCOL's independence somewhat reduces the risk of the same kind of corruption that plagues many other electrification agencies. Finally, it relies on an extensive nongovernmental organization network to verify that providers such as Grameen provide quality services (NREL 2015, 10), thus contributing to local accountability.

In sum, Bangladesh's rural electrification program has two prongs. One program is highly inefficient, prone to theft, underfunded, and unable to improve electrification rates. The other program performs much better but unfortunately has operated on a small scale for a long time. The REB is cost-effective, but it remains an open question whether it can scale up its operations to achieve universal electrification. In a worrying development, the REB's performance started to decline in the mid-2000s. The reason, Taniguchi and Kaneko (2009) argue, is politics. The very success of the REB makes it an attractive target for political capture. This feedback loop underscores the critical fact that the REB, a program that is well designed to tackle capacity problems, operates within a country that otherwise faces enormous challenges pertaining to threats to capacity building, most notably corruption.

Indonesia: The Limits of Democratization and Decentralization

Indonesia is the largest country in Southeast Asia and the fourth most populous country in the world, with more than 250 million inhabitants. The country is not contiguous; it comprises thousands of islands, though nearly 60 percent of its population lives on the main island of Java, making it the world's most populated island. In 1971, a few years after the authoritarian government of Suharto came to power, the installed generation capacity was only around 532 megawatts, with two-thirds located in Java, and the rural electrification rate remained at 1.8 percent in the mid-1970s (Kwon 2006). The electricity service during this time was unreliable and had high transmission and distribution losses (IDA 1972). Given the presence of water barriers, the location of electricity generation facilities and extension of the national grid have proven a considerable challenge to the government. Rural electrification rates increased to nearly 60 percent in 1996 (Kwon 2006), and the government's target is to reach 97.5 percent at the end of 2019. However, this still leaves 6.5 million people without access to the grid. This number could be higher since the state electricity firm, Perusahaan Listrik Negara (PLN), considers a village electrified even if there is just one distribution connection to the power grid (Economist Intelligence Unit 2014). In 2010, the generation capacity was just 29 gigawatts or around 0.12 kilowatt-hour per capita, among the lowest in the region.

Indonesia's slow progress in rural electrification and modern cooking fuels is particularly striking, given that the country has been relatively wealthy for a long time. In 1980, the country's GDP per capita was US$556 (see table 5.1), much higher than that of China and Vietnam, which have made tremendous progress in rural electrification. Conventional explanations like geography and resource availability also do not explain the variation in electrification rates in Indonesia. While Java is the most populated island in the country and also has a higher electrification rate relative to other provinces overall, not all parts of the province are well connected to the electricity grid. For example, Banten in western Java has a lower electrification level than Bengkulu on the island of Sumatra. Although the Indonesian archipelagic structure makes electrification an arduous task, population density alone does not adequately explain the variation in electrification levels in the country. In addition, resource availability is not sufficient to explain the slow growth of electrification rates in Indonesia, a country with abundant fossil fuel resources. Variation in resources for power generation within the country has only limited explanatory power. For example, South Kalimantan is one of the places with the highest coal production in the country; however, its electrification levels lag behind places on the island of Java. Hence, the availability of resources does not automatically mean higher electrification rates.

To understand the reasons behind why Indonesia has made only limited progress in rural electrification rates, we turn to our theoretical argument. Recall that high government interest is necessary to have any success in alleviating energy poverty, and having high institutional capacity and and high local accountability will improve the chances of success. In the case of Indonesia, we show that government interest was historically low for both rural electrification (until 2003) and cooking (until 2007). Though there has been a renewed interest in improving electrification rates over the past decade, the sector still lags behind in institutional capacity and local accountability. For much for the past six decades, PLN's financial stability has been dire, and coordination between different government ministries responsible for rural electrification is still low. Local accountability has increased over the past decade with the decentralization process, but regional authorities are still constrained when it comes to having responsibility over electricity distribution.

Government interest in rural electrification was always low under Suharto because of the centralized and authoritarian nature of his rule, along with the fact that his political survival was never really threatened until the Asian financial crisis (Vatikiotis 1999). Compared to the authoritarian governments in China and Vietnam, Suharto had a longer time horizon (Geddes 1999), allowing him to sustain lower levels of government interest in electrification. Early government interest in rural electrification began in the late 1950s after the electricity sector was nationalized under military law (McCawley 1970). The nascent program was not the result of a concerted effort to improve electrification rates, but an ad hoc setup with the assistance of different foreign partners.

This caused major maintenance problems over the next two decades with very little coordination between these different projects (McCawley 1978). During the 1970s, the government explored setting up microhydro projects to improve rural electrification rates. However, there was little clarity on the government's electrification policy, specifically on the roles and responsibilities of the PLN and different government units. Unlike other developing countries like India, which had set up the Rural Electrification Corporation by this time, there was no equivalent organization set up by the Suharto government. On the main island of Java alone, rural electrification policy in the late 1970s involved scattered distribution systems with power produced from diesel and hydroplants and financed by different levels of government and with a mixture of private and public operators (McCawley 1978). There was no clear pricing and connection policy, and the initial connection charges were very high, so only wealthy households in the rural areas benefited from these programs.

The collapse of the Suharto government in 1998 and the onset of democracy did not immediately change things; in fact, government interest took a turn for the worse with the disbandment of the PLN's rural electrification division in 2001 (World Bank 2005). It was only with the decentralization initiative that began in 2001 (Hofman and Kaiser 2006) that government interest in rural electrification made some headway. Decentralization allowed many districts to elect their mayors, who in turn pressured the national government to invest in rural electrification infrastructure and regulation. Over the past decade, the national government has passed a number of regulations that affect the electricity sector. These have mainly revolved around setting electrification targets, allowing private players to sell directly to end users and incentivizing the use of renewable energy.

Institutional capacity in the Indonesian electricity sector has revolved around the PLN, the organization responsible for the generation, transmission, and distribution of electricity. While this brought some much-needed capacity to the sector, the supervision of the PLN moved between different government ministries before finally coming under the Public Works and Electric Power in 1972 (IDA 1972). Moreover, the PLN was in dire financial situation in the 1960s, with costs being a huge problem. The costs of running a state electrification agency far exceeded revenues, especially when the government sought to control prices (McCawley 1970). Using money from the International Development Association, the PLN embarked on a new charter that gave it control over all dimensions of the electricity sector but with limited autonomy over prices, which still had to be approved by the Suharto government (IDA 1972). As of 1972, there were considerable arrears in what the government had to pay PLN. Foreign investors like the World Bank already had concerns about PLN's future growth, especially since the government had yet to substantially finance the PLN for any rural electrification expansion.

The PLN's financial viability was a concern well into in the late 1980s, along with other issues like managerial resources. The extension of the grid mainly benefited villages in the vicinity of around 3 kilometers of the grid, and all government subsidies to the PLN were used to keep prices down for rich customers rather than finance any rural expansion (Munasinghe 1988; McCawley 1978). During the 1980s, the institutional framework for rural electrification in Indonesia was based on the Ministry of Mines and Energy (which oversaw PLN) and the Ministry of Cooperatives, but there was very little coordination between the two ministries (Munasinghe 1988). Overall, institutional capacity in Indonesia has always been low and continues to be so today.

Indonesia has been a centralized state for much of its postcolonial history. During his tenure of more than thirty years, Suharto held an iron grip over almost all parts of the economy. All major regulations about rural electrification, including budgeting and finance, were decided in Jakarta and then filtered down to the rest of the provinces and outer islands. Little effort was made to identify the electrification needs of the population, thus limiting local accountability and bottom-up dynamics. The rural populace could also do little to signal their electrification needs to Jakarta or punish their local politicians for lack of effort since they were primarily selected by top military officials in the capital.

After democratization in the late 1990s, decentralization was introduced in 2001, with many districts now able to elect local leaders, most significantly their mayors (Hofman and Kaiser 2006). This gave the regional authorities some autonomy to focus on rural electrification expansion projects, engage the private sector directly for implementation, and channel subsidies to households that needed them the most (Differ Group 2012). While this policy change has largely proven useful, the rapid decentralization process led to the creation of new provinces and districts, and local authorities still largely depend on PLN for electricity generation and distribution. In other words, decentralization has increased local accountability in Indonesia over the past decade, but much more could be done. There are, in particular, no mechanisms in place that would assess local needs, engage local communities, and feed this information back up to local, regional, and national policymakers.

Since the election of Joko Widodo as the president of Indonesia in October 2014, there has been a pledge to improve archaic regulations and remove investment barriers. Specifically, he has spoken of "shaking up" the PLN, which continues to have a monopoly over electricity distribution.[9] These efforts to increase private participation in the electricity sector are clearly helpful, especially after the Asian financial crisis, but the current government's focus is still on improving electricity generation and less on the expansion of rural electrification. A coordinated national program for rural electrification is still missing from the government's agenda. The institutional capacity of the PLN and related government ministries also needs much improvement: rural electrification in Indonesia has usually come under different ministries with little coordination between various programs. Moreover, a decentralized electricity distribution model where a local-government-owned company is responsible for electricity distribution within its jurisdiction whereas the PLN continues to be responsible for electricity generation and transmission would be useful (World Bank 2005). Giving regional authorities the incentive to set tariffs and collect revenue will ensure that local leaders are incentivized to improve energy access in the country.

The government has shown little interest in improving access to clean cooking fuels. As in the case of rural electrification, this was mainly due to the fact that the Suharto government had very little incentive. Moreover, his government had to dole out huge kerosene subsidies to artificially lower prices—so much so that in 2006, kerosene subsidy levels were around 57 percent of the country's total petroleum subsidy (PT Pertamina and World LP Gas Association 2015; Asian Development Bank 2015). It was only in 2007 that the government put in place a program to incentivize users to switch from kerosene to LPG (Budya and Arofat 2011). Within a couple of years of the start of the program, the government had distributed over 44 million conversion packages consisting of an LPG stove, cylinder, first-gas fill, and accessories in over fifteen provinces (Andadari, Mulder, and Rietveld 2014). A recent evaluation of this program concluded that it was effective in making people switch to LPG, although there was some fuel stacking involving both electricity and traditional biomass (Andadari, Mulder, and Rietveld 2014; Soo et al. 2015).

This is clearly a step into the right direction, but about 25 million households still continue to use traditional biomass as their cooking fuel (IBRD

2013), and the viability of this initiative remains unclear as there is no guarantee that households, especially in rural areas, have the necessary LPG access to continue using the cleaner fuel in their cookstoves. More recently, the Indonesian government has teamed up with the World Bank to implement a Clean Stove Initiative that aims to promote clean cooking solutions for the entire country (IBRD 2013). In short, government interest in encouraging households to use LPG has increased in the past decade, but there is still no coordinated national plan to make this sustainable in the future. This is key if the country is to make progress in improving energy access in the coming decades.

In sum, rural electrification levels in Indonesia remain low relative to what they should be, considering the country's wealth and abundant fossil fuel resources, as well as the performance of much poorer countries elsewhere in developing Asia. The reasons for this have, until very recently, revolved around lukewarm government interest toward rural electrification. While government interest has grown recently thanks to the decentralization that followed democratization, the effect on rural electrification rates has only recently begun to show, with rates of 82 percent in 2016 (IEA 2017). Even if government interest for rural electrification continues to improve in the near future, institutional capacity continues to be a constraint. With no dedicated ministry to coordinate rural electrification, existing processes are inefficient and ineffective. Local accountability has also been poor during the Suharto period in Indonesia, and notwithstanding that democratization and decentralization have allowed for more local voices, much of the political elite structure that existed during the Suharto era continues to the present day. Until such time as institutional capacity and local accountability improve, rural electrification rates in Indonesia are bound to lag behind other countries with similar fundamentals.

Understanding Barriers to Energy Access in Africa: Nigeria and Kenya

Africa is the final frontier for energy poverty eradication. By examining the cases of Nigeria and Kenya, we shed light on why energy access has presented such a challenge for African governments and societies. In both cases, access to electricity and adoption of clean cooking fuels remain disappointing. The policy failure is particularly troubling in Nigeria, where abundant domestic energy resources would predict a high level of access in terms of both electricity and LPG.

Nigeria: Persistent Energy Poverty Despite Democratization

As the "Giant of Africa," Nigeria's population of 185 million makes it the most populous country on the continent and the seventh largest in the world. Its GDP in 2016 was US$457 billion (in 2010 US dollars), which means that it is Africa's largest economy, well ahead of South Africa. As an economic powerhouse, it accounts for almost a third of economic output in sub-Saharan Africa and saw growth rates consistently above 4 percent over the past decade.[10] Due to its sheer size and economic importance for the region alone, Nigeria would surely be an interesting case to study in its own right. However, what makes it even more interesting, specifically given our focus on energy poverty, is that Nigeria is among the countries with the lowest electricity-generating capacity per capita around the world. Its generating capacity in 2012 was as low as 6,090 megawatts—less than one-seventh of South Africa's capacity or about equivalent to the capacity of Cuba, a country with 11 million inhabitants and one-tenth of Nigeria's economic production (EIA 2015). Brazil, a country of similar size in terms of population, generates about twenty-four times the electricity that Nigeria does. In 2008, aggregate electrification rates remained at 50 percent, and outside urban areas at about 35 percent (Oseni 2012, 993–994). While total electrification stood at 61 percent in 2016, rural electrification remained low at 34 percent (IEA 2017). The track record for modern cooking fuels is even worse: 80 percent of households use firewood, and less than 1 percent can afford the use of gas for cooking (Oseni 2012, 993). Nigeria has thus largely failed to lift its population out of energy poverty. Why is energy poverty so persistent in a relatively wealthy country with abundant fossil fuels?

To date, Nigeria produces 80 percent of its generating capacity from fossil fuels and 20 percent from hydroelectric power. While peak demand has seen a rampant increase over the years, power supply falls short of demand by almost 50 percent (Oseni 2011). In fact, electricity access from public generation decreased from 47 percent to about 40 percent in 2008, generating the largest supply-demand gap in the world (Oseni 2012). This failure seems particularly ironic in light of Nigeria's vast oil reserves; it is Africa's largest oil producer and the fifth largest exporter worldwide of LNG (EIA

2015). Moreover, Nigeria's transition toward democratization in the late 1990s seems to have done little to mitigate energy poverty.[11]

Energy poverty in Nigeria challenges explanations that identify resource endowments or democratic institutions as the key determinants for alleviating energy poverty. Oil was discovered in Nigeria in the mid-1950s, and since then there has been a steady increase in national production. However, this resource abundance has not translated into a corresponding increase in GDP or electrification rates. About 98 percent of its crude oil is exported (Oseni 2012), and much of what is used domestically is lost to waste and corruption (Sala-i-Martin and Subramanian 2013). The Nigerian government, of course, benefits from oil revenues, which account for more than two-thirds of government revenue.[12]

However, large amounts of oil revenues do not end up in the government's coffers. Recently, *BBC News* reported that the Nigerian National Petroleum Corporation (NNPC) failed to pay US$16 billion to the national government, which accounts for about a third of the government's annual oil revenues.[13] While shocking, this is an accurate picture of daily Nigerian politics. A PricewaterhouseCoopers audit, in a separate instance, documents another US$19 billion that were kept by the NNPC from January 2012 to July 2013 alone.[14] A Reuters report suggests that half of the country's oil revenue from 2012 to mid-2015 never reached the federal government's balance sheets.[15] "Nigeria's oil reserves should have been a blessing for Nigeria to be used to build infrastructure and invest in social services. Instead, it has been a curse, a lubricant that has produced massive corruption and dysfunctional governments," thus undermining the explanatory power of resource availability for rural electrification and cooking fuel access.[16] Resources are there, but they are not used to help Nigeria's energy-poor.

Democracy has been not very helpful either. Nigeria has been under authoritarian rule for most of its postindependence history; it was only in 1999 that the country transitioned to democracy. Though democratic consolidation is still an ongoing process, democratization has not boosted electrification rates or access to modern cooking fuels in the previous fifteen years. In sum, conventional explanations fail to explain the energy access situation in Nigeria, and we instead revert to our theoretical argument to provide better insight.

We have argued that in order to find an appropriate solution to energy poverty, we need to have high government interest, good institutional capacity, and strong local accountability. In the case of Nigeria, we can show that only the first condition, government interest to improve electricity access, is met, at least on paper. Nigeria's complete failure to increase electrification rates should instead be attributed to limited local accountability and low levels of institutional capacity, as evidenced by the inefficiency and corruption of the National Electric Power Authority (NEPA), Nigeria's power sector regulator.

Faced with low electricity access, the first democratically elected government, under President Olusegun Obasanjo, who took office in 1999, set out to tackle energy poverty head-on. During the inauguration ceremony, the minister of power and steel announced what became ultimately known as the "power creed" (Olukoju 2004, 61), aiming to cut power failure by almost 100 percent within two years. Governmental rhetoric aside, during its first two years in office, Nigeria's federal government shored up funding for the NEPA to record levels. It disbursed 22 billion naira (about US$1 billion) in 2000 and more than 50 billion naira (more than US$2 billion) in 2001, more than a ten- and twentyfold increase from 1999 levels. Additionally, the federal government in Lagos signed several significant memoranda of understanding with private investors such as Shell, AGIP (Azienda Generale Italiana Petroli, later bought by Eni), and Eagle Energy to generate electric power from Nigeria's massive resources of natural gas (Olukoju 2004, 68).

In a recent interview, former president Obasanjo made clear what his main motivation was at the time: "Our emphasis should be on energy and infrastructure. If we get those two correct, we will particularly generate employment."[17] Energy access is thus considered important not only in its own right, but also for creating employment and promoting economic growth. In the words of Ikeme and Ebohon (2005, 1214), "for a developing country like Nigeria where poverty is endemic and rapidly unfolding into new forms and dimensions, sufficient production and consumption of electric energy has become the preoccupation of the government because of the constraints it poses to industrial capacity utilisation and employment-creating opportunities." Government interest in electrification is hence primarily a strategy to fight economic poverty. With 70 percent of the country's population living in rural areas and at least half the states having electrification rates below 50 percent (Ohiare 2015, 2), electricity access poses a problem to all states. As winning presidential elections requires support of two-thirds of Nigeria's twenty-six states, pushing for rural electrification is a politically attractive policy, especially after Nigeria became democratic in 1999. While Nigerian politics are typically entrenched in religion, race, and ethnicity, putting personalities at the forefront, the latest presidential elections, in 2015, were also fought over concerns of security and employment.[18] Appealing to rural voters becomes important in a nation in which almost every second person is living in poverty, and promising energy access, as Muhammadu Buhari did and won the presidency, surely builds electoral support.[19] In Nigeria, government interest in rural electrification goes hand in hand with an economic development strategy.

All of these efforts were mostly in vain. Olukoju (2004, 66) summarizes, "Despite the onset of democratic rule and the federal government's declared commitment to ensuring a steady power supply, no significant change has taken place in the power sector." The situation even worsened over time. An EIA (2015) country report on Nigeria finds that in 2011, Nigeria was the country with the largest gap between electricity demand and supply worldwide. This begs the question: Why is energy poverty so persistent in Nigeria despite high government interest at the federal level?

The explanation for Nigeria's continued failure to provide electricity lies with the inefficiency and mismanagement of the country's electricity regulator, NEPA. It was founded in April 1972 as a merger of two other regulatory bodies, the Electricity Corporation of Nigeria (ECN) and the Niger Dams Authority (NDA). ECN was in charge of generating and supplying power, while NDA was responsible for dam construction and the generation of hydroelectricity, mainly in the Niger delta; generated power was then sold to ECN for feeding it into the national grid. The main purpose of centralizing electricity services in a single authority was to simplify financial and organizational structures and use available resources much more efficiently (Okoro and Chikuni 2007).

Centralization, however, did not help alleviate energy poverty. Rather, the state monopoly in generation, transmission, and distribution proved unable to keep up with the quickly rising energy demand. Peak demand increased tenfold from 385 megawatts in 1973 to more than 3,000 megawatts in 1996 (Ikeme and Ebohon 2005), while stifled productivity, low accountability, and lacking investments in both generation and infrastructure are identified as primary obstacles to expanding power supply. Prior to reforming electricity markets, electricity rates were kept at artificially low prices to, at least in principle, grant access to the country's poor. Yet with rates covering only about half of long-run generation costs, investments stalled. Similarly, payments are collected irregularly, and Ikeme and Ebohon (2005) argue that only 50 percent of delivered electricity is paid for. (In comparison, we saw that Bangladesh's Rural Electrification Board collects over 95 percent of the money it is due.) Technology to transform primary energy into usable electricity is old and outdated, and its works poorly. Oseni (2011) finds, for example, that despite a generating capacity of about 6,000 megawatts in postreform years after 2005, only 40 percent of this produced electricity can reach Nigeria's households due to malfunctioning distribution infrastructure and faulty transmission equipment.

Poor maintenance and lacking investments aside, vandalism of equipment and corruption during implementation add to the problems of Nigeria's electricity sector and speak strongly to problems of institutional capacity. Destruction of equipment and infrastructure is particularly prevalent in the Niger delta, where warlords control most of the country's oil reserves (Okoro and Chikuni 2007). Attacks committed by local rebel groups, such as the Niger Delta Avengers, the Joint Niger Delta Liberation Force, or the Ultimate Warriors of Niger Delta have devastating effects. Destruction of Eni- and Shell-owned pipelines and explosions on two of Chevron's oil fields in May and June 2016 brought down Nigeria's oil production by almost 40 percent from 2.2 million to about 1.4 million barrels a day, making Nigeria only the second largest oil producer in Africa behind Angola, at least temporarily.[20]

Bribery and connivance of NEPA officials, "who derive gain from the misery of consumers and from the public purse" (Olukoju 2004, 58), are common. NEPA officials can, for instance, delay repairs or replacement of malfunctioning equipment until households pay for these services. These practices undermine local accountability at the operational and political levels, enticing households to steal power, tamper with meters, or illegally reconnect to the grid in other ways. The severity of NEPA's inefficiency and malpractice is so flagrant that during the African Nations Cup soccer tournament in January 2002, twelve fans cheering for Nigeria's Super Eagles were killed when a TV set exploded after NEPA restored power and overloaded the circuit (Olukoju 2004, 59). These are by no means the only victims of NEPA's failure to reliably provide electricity as many households in urban areas invested in petrol-powered or diesel-operated electricity sets, while poorer families rely on candles and kerosene lamps, with resulting fire and health hazards. Lack in institutional capacity critically links to Nigeria's energy poverty, making Nigerians joke that the acronym NEPA stands for "Never Expect Power Always."

NEPA not only lacks institutional capacity; it is also deficient in creating institutional structures for local accountability. Olukoju (2004, 55) argues that almost by design, equipping NEPA with monopolistic powers undermines local accountability. Nigeria's regulatory agency is so removed from oversight that it is unlikely to suffer consequences from malpractice and nonperformance. In a country that has seen thirty years of military rule and where corruption is ubiquitous, there is little fertile ground for local accountability (Olukoju 2004, 56). Oseni (2011, 4772) goes so far as to say that "corruption is a disease that has eaten deep in the bone marrow of many Nigerians, especially in the public sector." Accountability requires trust, oversight, and information transmission, none of which is currently a reality in Nigerian energy politics. In a recent study, Ikejemba and Schuur (2016) find that solar projects in Nigeria are mostly doomed to fail because the projects are not adequately integrated into local communities. In essence, electricity demand and local needs of rural populations are ignored, as accountability is absent. Communication between those wanting electricity access and those providing fails.[21]

These failures started attracting attention by the country's leadership. Even though the latest election in March 2015, which resulted in Muhammadu Buhari's win, were largely run on a security platform over fear of Boko Haram's terror in the northeastern part of the country (Buhari is a former military leader), the lack of electricity access was a salient issue among Nigeria's population.[22] The Electric Power Sector Reform Act had been enacted in 2005 to privatize large parts of Nigeria's electricity market. NEPA was transformed into the Power Holding Company of Nigeria (PHCN) and aimed to unbundle electricity services and open markets for private investment. Despite these efforts, a substantial restructuring of the market did not take effect by 2010, and first steps toward privatizing generation and distribution were taken only in 2013, more than eight years after the legislation was written into law (EIA 2015). Reform success is slow, as state officials cling on to their power monopoly and delay speedy reform (Oseni 2011). PHCN and its subsequent eighteen subsidiary companies could not improve electricity access by much, if at all. Public generation capacity saw a net decrease in electricity provision by 7 percentage points in 2008 to the year prior (Oseni 2012), only nineteen of twenty-nine installed generating plants are operating (Oseni 2011), and Nigeria's enormous potential from renewable resources remains largely untapped (Oyedepo 2012; Ikeme and Ebohon 2005). Solar power projects, for example, suffer from a lack in awareness by local governments, implementing agencies, and project developers who do not know how to deliver these projects (Ikejemba and Schuur 2016). Poor governance and bad management skills remain problems to this day.

The economic cost of these management failures is high. Even after the reform, MTN, a South African mobile phone provider, installed about 6,000 diesel generators to supply base electricity and ensure its business operations. The associated costs to MTN, according to Oyedepo (2012), amount to about US$5.5 million annually in diesel fuel alone. MTN is no exception: 97 percent of firms own captive electric generation plants (Ikeme and Ebohon, 2005). More generally, low rates of institutional capacity and local accountability have compromised the quality of electricity service in Nigeria. It is estimated that between 1999 and 2007 alone, more than US$16 billion were sunk in the energy sector without seeing major improvements in power output, which suggests that most of this money was channeled toward corruption and bribery.[23]

There may, however, be some hope. Buhari, Nigeria's president who peacefully took over office from Goodluck Jonathan of the long-ruling, dominant People's Democratic Party, announced that he would tackle corruption vigorously. In a country that has presumably lost US$4,400 billion to corruption since its independence, these efforts would have important implications for rural electrification too.[24]

The situation for cooking seems even worse, and researchers often refer to Nigeria's "cooking energy crisis" (Anozie et al. 2007, 1283). Access to modern cooking fuels ranks low on the government's agenda. While documenting an absence of evidence is challenging, no Nigerian government has ever put forward a strategy on access to modern cooking fuels in a similar way as they promoted rural electrification. Again, this is consistent with our argument that ties government interest in electrification to identifiable economic benefits, for both industrial production and agricultural renewal. Neither is the case, though, in the context of cooking.

About 95 percent of the country's entire firewood consumption is for household cooking (Oyedepo 2012), and 80 percent of households use firewood as primary cooking fuel (Oseni 2012, 993). This share has even increased recently, while the share of households using kerosene dropped from 22.9 percent in 2007 to 18.5 percent in 2008 because of high market prices for kerosene, which the country's poor cannot afford. While there is variation between urban and rural areas in terms of kerosene use, with almost 90 percent of households using it in Lagos, the country's largest city and a major financial center in Africa, but less than 2 percent using it in the more rural northeastern states of Bauchi or Jigawa, kerosene is by no means a modern and safe cooking fuel. As few as 0.7 percent of households in 2007 used natural gas, which is not surprising "given that many Nigerians are poor and cannot afford the cost of gas for cooking" (Oseni 2012, 993). Although the Global Alliance for Clean Cookstoves reports some uptick in the use of efficient cookstoves in Nigeria, according to its database, only about 30,000 such cookstoves have been distributed so far, mostly as small-scale Programme of Activities projects under the Clean Development Mechanism framework.[25]

In a country where 60 percent of the population lives on less than US$2 per day, the primary obstacles to better access to modern cooking fuels are income and lack of capital. The high upfront costs make a transition toward modern cooking fuels difficult. That said, the recently launched National Clean Cooking Programme is intended to address this issue. In cooperation with the Global Alliance for Clean Cookstoves and with financial support from the Nigeria Infrastructure Advisory Facility, Nigeria's Federal Ministry of Environment started this initiative as part of the national Renewable Energy Programme, with the ambitious goal of facilitating the adoption of clean and energy-efficient cookstoves in 17.5 million households across the country by 2020.[26] Whether this program, launched in April 2014, is going to be successful in transforming cooking in Nigeria is for the future to tell.

Finally, there is no prominent role for new technology yet in solving Nigeria's problem of energy poverty, but the Nigerian government has been adamant about facilitating public-private partnerships in order to both boost foreign direct investments and acquire technological know-how (Olukoju 2004). This strategy may prove particularly useful for unlocking electricity generation from renewable sources. To date, it remains to be seen to what extent technology will mitigate energy poverty.

Nigeria is a particularly tough case for theories that seek to explain energy poverty. Nigeria is rich in oil and natural gas and has seen a democratic renewal, but despite these favorable circumstances, it does poorly in rural electrification. Although the federal government has a strong interest in addressing the problem of low electrification rates, mainly to promote economic development, substantial roadblocks remain. If anything, the Giant of Africa's record in access to modern cooking fuels is even worse.

Energy poverty in Nigeria is persistent even after power sector reforms were enacted in 2005. Consistent with our expectations, we find strong evidence for a lack of institutional capacity with the country's dysfunctional electricity regulator, whose monopolistic powers also undermine local accountability. Absent any oversight, the regulator even appears to have aggravated the problem, taking advantage of customers' intermittent power supply. Government interest, institutional capacity, and local accountability are once more key to understanding persistent energy poverty in Nigeria.

Kenya: A Bright Future Ahead?

Energy poverty in Kenya is a major hindrance to the country's economic growth and development. In 2010, only 16 percent of the total population, and about 2 percent of its rural population, had access to electricity (IEA 2010a).[27] Heavy reliance on hydropower in a country that experiences many droughts means that it is often vulnerable to power shortages (Kapika and Eberhard 2013, 27). The situation is made worse by high transmission and distribution losses, although the situation has improved in recent years (Kapika and Eberhard 2013, 25). Even in places with high population density and ample grid coverage, electrification rates are low (Lee et al. 2014). At the same time, Kenya has been a pioneer in off-grid solar technology since the 1970s. On the cooking side, there is little access to clean fuels, and firewood is by far the most prevalent energy input for cooking (Nyoike 2002). According to these data, Kenya must clearly be classified as an energy-poor country. What can explain these low rural electrification rates and low access to clean cooking fuels, despite changes for the better lately?

Conventional explanations are inadequate to explain this situation. For example, geography does not explain why Kenyan rural electrification levels have long been so low. Kenya's geography is a diverse mix of plains, hills, and forest region. This has contributed to varied population densities throughout the country. While places like Nairobi and the Central, Nyanza, and Western Provinces of the country have a large number of residents, there are far fewer on the coast and in the eastern parts of the country. Geography therefore played an adverse role for one part of the population. However, rural electrification rates have been fairly poor even in places with high population density. In a recent study, Lee et al. (2014) find that electrification levels are still low in places with high population density. Thus, this is not an adequate explanation of energy access in the country. Neither can income alone explain Kenya's energy poverty. Kenya was (and remains) quite poor, but its situation in the 1970s and 1980s was fairly comparable to Ghana's. As we saw in chapter 5, Ghana managed to deal with its energy issues despite limited resources.

To explain poor energy access in Kenya, we turn to our theory that emphasizes government interest, institutional capacity, and local accountability. We start with government interest, which for a long time was low in comparison to other countries. The Kenyan government's initial efforts to deal with rural electrification date back to 1973, when it launched the Rural Electrification Program (REP). The REP struggled to mitigate energy poverty. Partly this was because it focused only on subsidizing the cost of electrification (Lee et al. 2014). But the problem runs deeper: Kenya in general showed few signs of strong government interest, at least in the initial years after the country's independence. Bhattacharyya (2013, 147) notes that "Kenya can be considered as a rural electrification paradox" because it has struggled with energy poverty even though policies to tackle it have been in place for many decades. In his view, the poor record on energy poverty is due to a lack of "strong government commitment."

Although explaining this lack of interest is difficult, two drivers appear important. First, successive governments were discouraged by the high cost of rural electrification. According to Lee et al. (2016, 27), the "cost of grid expansion was prohibitively high and there was a general perception that demand for energy in rural areas was too low to be financially viable." The problem was worsened by declining income among rural farmers (Hornsby 2013, 368). However, a similar argument can be made for many other countries, and cost alone is not a sufficient explanation for the lack of government interest.

Perhaps more fundamental, Kenya exhibited a high degree of urban and ethnic bias. Jomo Kenyatta, the country's first president, was regularly accused of favoring the Kikuyu people, who were geographically situated in the center of the country. Areas populated by Kikuyu people benefited from higher electrification rates (Nyukuri 1997, 9). Other areas were left at a disadvantage. Places inhabited by Luos, for instance, "suffered severe repression and neglect" (Nyukuri 1997, 9). This bias was reflected in energy policy; plans to construct power plants in non-Kikuyu regions were canceled during Kenyatta's rise to power, showing the strong bias toward the center.

President Daniel arap Moi, Kenyatta's successor, who ruled Kenya between 1978 and 2002, faced similar incentives, even though he was not part of the Kikuyu. He had to ensure that the population of its urban centers remained content. Ogot and Ochieng (1996, 125) note that "the state spent larger sums per capita on housing, schools, health centres and roads in the urban areas. Virtually all new industrial development was centred in existing urban areas. The growth of the urban population, the economic significance of the urban areas and the potential political and social volatility of such large centres as Nairobi and Mombasa led to further examples of urban bias." In sum, electrifying rural settlements represented a high financial cost for a small political benefit.

The situation changed gradually in the late 1990s and early 2000s, coinciding with the end of Daniel arap Moi's tenure and the democratization of the country. In 1997, the government decided to enact the Electric Power Act and establish the Rural Electrification Authority (REA) to put in place a plan to provide electricity to rural parts of the country. A new energy act that aimed to accelerate rural electrification under the REA was passed in 2006 (Kiplagat, Wang, and Li 2011). It also established the Rural Electrification Programme Fund, whose aim was to raise money for the government's electrification initiatives. Though the REA mandate includes supporting the use of renewable energy, it has until recently mainly focused on hydropower and geothermal energy. The REA was also tasked to involve private players, but energy markets in Kenya are still heavily regulated, and there remain significant barriers to private market entry (Abdullah and Markandya, 2009).

The prospects of the REA—and the lack of success of the REP—are related to lacking institutional capacity, which remained low until the end of Moi's regime. The Kenya Power Company, founded in 1954, was originally in charge of Kenya's entire electricity supply chain. Starting in 1997, the power sector underwent a series of reforms. One result was the unbundling of the public energy sector. This resulted in the creation of the Kenya Electricity Generating Company, KenGen, which has since been responsible for electricity generation. While KenGen holds a market share of 70 percent, leaving about a third of the market to privately owned, independent power producers, transmission and distribution infrastructure remains in public hands (Abdullah and Jeanty, 2011). The Kenya Power and Lighting Company (KPLC), founded in 1922, was now in charge of electricity transmission, distribution, and retail. To crown this market, the Kenyan government created the Energy Regulatory Commission (ERC) in 2006, another product of the series of institutional reforms started in 1997. Initially the government had empowered a relatively weak Electricity Regulatory Board. In 2006, it replaced it with the more powerful ERC, putting it in charge of energy regulation and monitoring the electricity market (Kapika and Eberhard 2013, 30).

The energy market setup compromised institutional capacity. For a long time, capacity was poor, and system losses were typically above 20 percent (Kapika and Eberhard 2013, 25). KPLC was, and remains, notorious for corruption, with contracts often allocated under the table; it is known for poor accounting and auditing (Hall 2007, 4). In addition, KPLC was also eager to shift the financial burden of its activities to other state-owned enterprises that, because of this, remained chronically underfunded (Karekezi and Mutiso 2000, 96). Worse, leaders such as Moi used KPLC as an extended arm of the state's surveillance apparatus (for instance, to gain entrance to people's homes) (Mueller 2008, 196). Similar problems have plagued the Rural Electrification Authorities (Abdullah and Markandya 2009, 6).

The reforms undertaken at the turn of the century seem to have improved the situation. Since then, system losses have decreased to about 15 percent (Kapika and Eberhard 2013, 25). To some degree, the reason is that the financial situation of the market has improved. Reforms unlocked private investment, with the development of a number of projects funded by both the private and the public sectors (Gratwick and Eberhard 2008). In addition, new regulations emphasize the need for better fee schedules. As

Kapika and Eberhard (2013, 39) note, "consumers will be expected to pay a tariff that at least covers the cost of generation." The capacity of the regulatory apparatus has also been improved. The ERC was established in 2006 to improve the energy market governance. As it turns out, the ERC is a fairly powerful entity (Kapika and Eberhard 2013, 31): it has the power to sue, it can raise money, and it can buy assets. Moreover, it is also an independent entity, which protects it from interference from both the government and the private sector. While its head is appointed by the government, the ERC has access to its own stream of revenues, which reduces the influence of Kenya's authorities (Kapika and Eberhard 2013, 34). Overall, we may note that Kenya long suffered from low levels of institutional capacity, but that recent reforms suggest that the problem has been acknowledged by a government that appears more interested than earlier in reducing energy poverty.

We now come to the third component of our theory: local accountability has generally been very low in Kenya, especially after the centralization of its government in the 1970s (Devas and Grant 2003). Recent reforms, however, suggest that the Kenyan government is seeking to improve this facet of the problem. The ERC, for example, has to submit both its books and its activity reports for external review (Kapika and Eberhard 2013, 35). This, however, is a fairly standard procedure. More important, individuals who feel that they have been unfairly treated by the ERC can appeal to the so-called energy tribunal, which can rule over process and policy. For instance, the tribunal could, at least in principle, invalidate rate orders. Furthermore, the REA has begun to adopt a more bottom-up approach that involves local leaders at various stages of an electrification project (Kenyan Ministry of Energy and Petroleum 2015).

Overall, then, the situation is summarized as follows. Government interest long remained fairly low. Capacity suffered, and there was little local accountability. Starting in the late 1990s, a series of reforms began to change that picture. An institution devoted to rural electrification was established, and the regulatory authorities were strengthened. Of course, the country remains poor, and corruption scandals suggest that Kenya has not yet found an effective institutional solution. However, the signs for the long run are now more positive than they used to be.

The prospects of rural grid electrification are closely related to another interesting development in Kenya: the success of off-grid solar technologies.

On a per capita basis, Kenya ranks among the first in terms of adoption of solar home systems (Martinot et al., 2002; Kiplagat, Wang, and Li, 2011). Ondraczek (2011, 12) estimates that about 320,000 solar home systems have been sold in Kenya since the early 1980s (for a population currently estimated at 48 million). He believes that more than 4 percent of rural households own such a device. Thus, Kenya's success with off-grid technologies actually rivals that of Bangladesh.

Why, then, have off-grid technologies been more successful than grid extension? In part, off-grid technologies have bypassed some of the obstacles that paralyzed the development of the grid. These technologies do not suffer as much from issues related to local accountability, making them a successful case of community-led solution to energy poverty (Chaurey and Kandpal, 2010). Highly relevant to our argument, Acker and Kammen (1996) identify an enabling local environment as *the* necessary condition for the proliferation of solar photovoltaic systems. Specifically, they claim that "programmes of this type [solar photovoltaic microgrids] require extensive local networks to implement, and are, by definition, not well managed on a top-down basis" (Acker and Kammen 1996, 109). Not only does access to microgrids increase incomes of rural populations by 20 to 70 percent, but cost recovery has also been proven to be much higher once microgrids were managed locally, so that energy services could be tailored to rural villagers' local needs and preferences (Abdullah and Jeanty 2011; Abdullah and Markandya 2009; Chaurey and Kandpal 2010). In their study of the effects of a microgrid, Kirubi et al. (2009, 1219) share a similar view and believe that systematic community involvement would be highly beneficial: "Such a move would raise awareness and provide critical public oversight and accountability. Such oversight is likely to happen because target communities, local elite, and politicians have the motivation and information to ensure that their villages are not short-changed and/or bypassed when their regions fall due for electrification." What is clearly working for off-grid solar systems could also work for grid extension. Abdullah and Jeanty (2011, 2981) suggest that the government of Kenya "through the local authorities, could subsidize a third of the connection cost with the rest being paid by the households" to increase electricity access not only through photovoltaic systems but also grid connections. This proposition is similar in spirit to Ghana's Self Help Electrification Programme in that involvement of local communities promotes local accountability and helps leverage the associated benefits from superior information transmission, mobilization of rural households, improved monitoring of equipment use, and more efficient resource allocation. In other words, local accountability has generally been low, but there have been some instances, especially with off-grid electrification, where local accountability has proven beneficial. This provides strong evidence in support of our theory: local accountability does matter and plays an important role in determining how effectively policies can address energy poverty. This bodes well with a recently published evaluation report of Kenya's REP, which concludes that "the government as a facilitator of the REP needs to be more transparent and accountable in its actions, so as to increase the efficacy of the electricity services, to both current users and potential consumers" (Abdullah and Markandya 2009, 35).

As is true for most other sub-Saharan African countries, Kenya relies heavily on firewood as the most prevalent cooking fuel, which accounts for more than 70 percent of energy demand (Nyoike 2002; Karekezi 2002). The death toll of using traditional biomass is around 15,000 Kenyans a year, mostly women and children. The health of almost 15 million people in Kenya is directly affected, according to an action plan by the Global Alliance for Clean Cookstoves (2013, 3). This action plan, which is implemented with the help of international partners, aims at distributing 7 million clean cookstoves by 2020 to alleviate energy poverty, especially in rural Kenya. While the use of LPG has been growing recently, now being used by about 5 percent of households, access to modern cooking fuels is limited to urban and rich parts of the country's 48 million population. Despite there being no program by the Kenyan government, there are several international initiatives—for example, the Global Alliance for Clean Cookstoves, Practical Action, and Programme of Activities (PoA) projects under the Clean Development Mechanism (CDM). Bilateral efforts, for example, by the United States, which support the distribution of clean cook stoves with US$125 million, complement the policies on the ground.[28]

Despite these extensive efforts, the uptake of efficient cookstoves remains slow. The small-scale CDM project, registered as PoA 5336, for example, reports that the adoption of modern cookstoves was lower than initially hoped, efficiency rates were less impressive due to tampering with the stoves, and many households continued to use their old stoves.[29] This is not an uncommon finding (Bansal, Saini, and Khatod 2013; Levine and

Cotterman 2012; Mobarak et al. 2012), which Sesan (2012) attributes to poverty. Even for cheap or free stoves, households need to buy fuel from markets that they may not be able to afford, calling for a more incremental transition toward modern cooking fuels. Another impediment for adopting efficient stoves among West Kochieng households, a community in Western Kenya, was that stoves were not designed to accommodate rural villagers' preferences (Sesan 2012, 202). Reflective of our theoretical argument, energy poverty can only be addressed once local needs are considered. Local accountability does not only entail adequate political representation and responsiveness by local decision makers; it also manifests itself in the form of accountability of international donors to local project "constituents" (Bailis et al. 2009; Lambe and Atteridge 2012). This is the more true for the case of a cooking revolution, as energy services for cooking are more differentiated products relative to providing electricity.

Overall, Kenya has had low electrification rates and poor access to modern cooking fuels in the past because of low levels of all three factors of our argument: government interest, institutional capacity, and local accountability. Technology can help alleviate energy poverty, but advanced technologies are often out of reach for developing countries. International support, perhaps through the World Bank as with Kenya's geothermal power generation, can successfully turn the tide as long as the national government is supportive. Government interest is thus once more critical. As the economic hub in East Africa, some observers even talk about an "energy revolution" in Kenya linked to the country's expansion in geothermal, hydroelectric, and solar power.[30] However, this example reminds us strongly that institutional capacity and local accountability matter too. The current off-grid solar progress that has been made in Kenya requires a sustainable institutionalized approach to ensure that the rural population is able to avail itself of these technologies at affordable prices. Moreover, after local preferences were taken seriously and management authority was delegated to local levels, solar photovoltaic systems saw high demand (Kirubi et al. 2009; Abdullah and Jeanty 2011; Abdullah and Markandya 2009; Chaurey and Kandpal 2010) and efficient cookstoves were adopted (Sesan 2012). With Kenya's new constitution since 2010, which devolves substantial power to forty-seven semiautonomous counties, the hope remains that this not only improves the people's political representation but also increases access to electricity and modern cooking fuels. Assuming our argument is correct, Kenya may have a bright energy future; the latest IEA (2017) data suggest nothing less than a major breakthrough with an increase in total electrification rates up from 18 percent in 2010 to 65 percent in 2016.

More Success in Clean Cooking Than in Rural Electrification: Senegal

Senegal is possibly the only country in the world in which access to cooking fuels is less of an issue than access to electricity. At first sight, it would appear as if Senegal falsified our general prediction that access to electricity occurs before access to modern cooking fuel. Upon closer inspection, however, the case reveals the power of government interest: successive governments cared little about providing electricity beyond the main industrial centers, whereas modern cooking fuels became a national priority because the country was threatened by desertification. The fact that it was able to create a functioning LPG market underscores the importance of political factors.

Senegal has so far failed to provide electricity to a vast share of its population. The situation was particularly bad until the late 1990s, as rural electrification stood at a paltry 5 percent in 1997, with few signs of improvement at the time (Mawhood and Gross 2014, 482). Government support was intermittent, and infrastructure in general lagged behind. The situation improved slightly in the late 1990s. In parallel to the country's transition to democracy, the new government under Abdoulaye Wade undertook a series of reforms designed to bolster electricity access, with a particular focus on rural communities. Still, progress has been slow. By 2010, only 54 percent of Senegal's households were connected to the grid (Mawhood and Gross 2014, 482). This low rate was mostly due to problems in the rural areas, where electrification rates had reached only about 24 percent. While overall electrification rates increased slightly to above 60 percent in 2016, every second Senegalese living in the countryside still lacks electricity access (IEA 2017).

In stark contrast to the electricity situation, modern cooking fuels have been widely adopted by consumers. Numbers vary, but it is generally accepted that between 20 and 40 percent of the population uses LPG, a staggering number even compared to other countries that have sought to promote LPG (Schlag and Zuzarte 2008; Brew-Hammond 2010). As we shall see, these successes stem less from a coherent long-term energy policy than from the urgency caused by deforestation. Until the 1970s, the fuel of choice for cooking was wood. This led to deforestation, which made the country vulnerable to desertification. To respond to this major risk, the government incentivized the use of butane (Schlag and Zuzarte 2008, 5).

With respect to electrification, then, Senegal represents a clear failure. Access is low, and there are few reasons to be optimistic about the near future. Increased public support in the late 1990s marked the beginning of a series of reforms, but the generally corrupt state of politics in Senegal seems to constitute an overwhelming barrier. But Senegal has been a leader in modern cooking fuels since an external shock, desertification, forced the government to intervene. Desertification increased the government's interest in reducing the use of biomass, and authorities gradually created institutions and markets to sustain the widespread use of LPG.

Traditional explanations such as geography, resources, and income are of limited use in understanding the outcome. Senegal is a small country without particular topological difficulties. It is also poor, but not poorer—at least until recently—than Ghana, which saw lots of success in rural electrification. Senegal has few energy resources, except for some hydropower in the Senegal river basin, but this lack of resources did not stop the government from developing a successful LPG program.

For most of Senegal's history, starting with its independence in 1960, government interest in rural electrification remained lukewarm at best. Unsurprisingly, this lack of interest coincides with the country's autocratic era (Galvan 2001). Despite the relatively benign leadership of Léopold Sédar Senghor's (1960–1981) and Abdou Diouf's (1981–2000) Socialist Party, Senegal remained solidly under the control of a single party. Electricity generation remained high enough to keep the economy afloat, but little progress was made to provide access to private households (Fall and Wamukonya 2003).

The situation changed marginally with Abdoulaye Wade's election in 2000. Wade's arrival coincided with an attempt to reform the electricity sector. Three main changes were announced in 1998 (Thiam 2010): privatization of Senegal's main electric company, SENELEC; the creation of ASER (Agence Sénégalaise d'Électrification Rurale), an agency devoted to rural electrification; and a program to promote the use of renewable energy. The last two reforms were specifically designed to solve the problem of rural electrification. ASER's goals were clarified in 2002 through the PASER (Plan d'Action Sénégalais d'Électrification Rurale), the Senegalese Rural Electrification Action Plan (Mawhood and Gross 2014, 481). Yet government interest remained unstable. Mawhood and Gross (2014, 486) argue that "[f]luctuating political support is considered to have been a major hindrance to PASER's development." Wade's tight grip on power possibly reduced his incentives to improve the sector, and corruption remained rampant.

Besides weak government leadership, the second critical issue is low capacity. Due to weak government interest, the electricity sector never fully overcame its systematic infrastructure shortcomings. For most of its history, Senegal suffered from poor infrastructures due to political meddling. Before the 1998 reforms, electrification was seen as a political tool to reward loyal supporters. In an interview, a government official states that "the selection of villages [for electrification] was viewed as inequitable, being based on proximity to the existing grid or political motivations" (Mawhood and Gross 2014, 482). Even the reforms did little to alleviate the problem. SENELEC, the country's main electricity producer, suffers from systemic deficits that paralyze its ability to invest in modern plants and new transmission lines (Sanoh et al. 2012). In fact, SENELEC faces a conundrum: it is highly dependent on foreign oil to power its thermal plants, which is costly, but it is politically unpopular to raise rates. In a report, the international organization IRENA (2012, 30) states, "the existing high cost of electricity to consumers means that it is impossible to finance further new connections through increased consumer tariffs and the scarcity of capital available to the government means that investments have to rely heavily on donors." Similar issues plague rural electrification programs such as ASER. Modibo Diop (ASER's former chief executive officer) and Pape Diallo (ASER's former chief financial officer) were each sentenced in 2013 to five years of prison for corruption.[31]

Finally, ASER has always suffered from a low level of accountability. The agency's leadership has often been chosen based on political considerations, with little regard to technical and professional expertise (Mawhood 2012). A number of observers "believe that some of ASER's staff—especially those at a senior level—were recruited to support the [then] current political regime rather than for their technical and professional merits" (Mawhood and Gross 2014, 485). As the government, ministries, and ASER battle and divert resources with a sense of impunity, accountability remains in short supply, and local stakeholders thus have little voice at any stage of the electrification process. The Programme Prioritaire de l'Électrification Rurale (Rural Electrification Priority Program), ASER's main plan under PASER, divided the country into ten regions, and the government then auctioned concession contracts to private actors, such as foreign utilities, with little input from the local population.

However, authorities seem to have recognized this problem. The Électrification Rurale d'Initiative Locale program (Local Initiative for Rural Electrification), a separate endeavor under PASER, sought to involve local communities more heavily in the process. Here local stakeholders must design a business and a feasibility plan to obtain funding (Contreras 2006, 8), but the program remains small and nascent. Ironically, while this particular program scores well on accountability, it requires so much input from local communities that, in the words of Mawhood (2012, 50),

there have been difficulties in establishing truly 'bottom-up' [Local Initiative for Rural Electrification] projects, with only a small number under preparation so far. It has been noted that rural communities do not necessarily have the organisational capacity to initiate local projects, and even those that do may be unaware of [the Local Initiative for Rural Electrification]'s existence.

The balance between high capacity provided by central authorities and strong accountability is sometimes difficult to find.

In recent years, tensions over the electricity problem have erupted. In 2011, a massive contestation movement against Wade was triggered by repeated electricity blackouts in Dakar. Exasperated, two journalists and two young artists launched the "Y'en a marre" movement (colloquially: "fed up"), which became particularly popular among the youth. This was the beginning of what some called the "electricity riots."[32] The political implications were not lost on Wade. In the 2012 elections, about one-sixth of the voting population were under twenty-three years old.[33] Despite last-minute attempts to woo the youth vote, Wade lost the elections to Macky Sall.

The contrast with the cooking fuel situation could hardly be starker. Over the past four decades, LPG became a dominant fuel, unlike most of the other countries that we have explored in this book. In the 1960s, charcoal and wood fuel were the primary energy sources for cooking (Fall et al. 2008). This presented the danger of deforestation and desertification due to the expansion of the nearby Sahel desert (Lazarus, Diallo, and Sokona

1994), making droughts much more dangerous and imperiling the stability of food production. Deforestation also destroys the natural habitat of various animals, further decreasing the country's food base (Sokona and Deme 2003). In sum, business as usual created a serious source of uncertainty and potential political risk for the government.

In the face of these significant threats, Senghor's government reacted by promoting the deployment of LPG and LPG-compatible stoves. In 1974, under a general "butanization" program, import duties on cookstoves were removed (Sokona and Deme 2003, 116). A few years later, the government started subsidizing LPG cylinders directly, though it occasionally reneged on this policy. Because of the relatively slow penetration of LPG, the government decided to boost subsidies again in 1987, and LPG use spread rapidly, though mostly in Dakar and other urban centers. Ten years later, as the state faced considerable fiscal issues, the government decided to reduce its support. Although LPG consumption decreased among the poorest, LPG had made a breakthrough. One study conducted in 2007 found that some neighborhoods outside Dakar had an LPG use rate of almost 100 percent (Fall et al. 2008). Urban areas in general are believed to have a use rate of about 70 percent (GNESD 2014, 40). Rural areas trail considerably, of course, but rates across the country are much higher than in most developing countries (Prasad 2008; Laan, Beaton, and Presta 2010).

Over the years, the government created an LPG market that is characterized by a level of capacity that clearly puts the electricity market to shame. The Ministry of Energy is advised by the National Committee for Hydrocarbons, whose expertise draws on specialists from energy, transport, finance, or trade (GNESD 2014, 42). Refining is done by Sengaz, a public company, and distribution is controlled by private multinationals such as Total. The entire system is thus one of relative balance between public and private interests in a way that is absent in the electricity market. The role of private interests is important. Initially the government directly subsidized small LPG cylinders (to encourage household consumption). Later, as these subsidies were scaled back, a lively private market developed to replace the government. Thus, local accountability now operated via competition, ensuring that prices remained affordable (Schlag and Zuzarte 2008, 11).

Even in LPG distribution, however, both local accountability and institutional capacity could be improved. Besides the usual issues related to high gas prices, the most immediate obstacle to a more thorough success of LPG comes from various weaknesses along the supply chain. The path from refineries to customers is a long one and involves private, public, and international actors. For instance, threats of strikes regularly disrupt the supply of LPG. The country has relatively few reserves, forcing households to rely on alternatives. In such situations, customers are powerless and accountability problems acute. Capacity is an issue too. Many wholesalers suffer from weak finances driven by the poverty of their customers and the general low reserves (GNESD 2014, 43). In addition to these problems, the successful provision of LPG requires well-functioning infrastructure, such as roads. Even suppliers that have trucks must occasionally rely on horse carts. Still, these problems pale in comparison to the remarkable progress Senegal has made (IPCC 2007, 719).

The Senegalese government has largely failed to solve the electricity access problem, but it has made progress in tackling the issue of access to modern cooking fuels. Far from casting doubt on our theory, we believe that the opposite is actually true: our theory helps us understand why cooking has been so successful while electricity failed to reach the poor. In particular, government interest is the key factor here. The success of the LPG program underscores how important a motivated government is. When the government sees little benefits from a policy, it can be expected to fail to provide the public good. However, when the government has an incentive to act (here, to protect the country against the perils of desertification), it seems as if no obstacles can really prevent it from doing so. This has important implications: most accounts of energy poverty argue that there are some fundamental barriers against affordable and abundant energy. For instance, energy poverty is often explained by pointing to the lack of natural resources. The case of Senegal shows the limits of this argument: even a poor country without energy resources such as Senegal can create a successful LPG program. The main obstacles are political.

Assessing the Cases: Determinants of Success and Failure

In this and the previous chapter, we have investigated eleven cases of success and failure. In these cases, strong government interest has emerged as the most important predictor of success in the eradication of energy poverty. Where government interest has been strong, institutional capacity and local accountability have contributed to rapid progress. These claims, of course, need to be considered in the context of geography, energy resources, and preexisting levels of economic development. Our evaluation of success and failure has been consistently based on prior expectations, which themselves are based on the three fundamental factors and any other relevant considerations. What makes cases such as Vietnam so remarkable is that their success cannot possibly be attributed to favorable preconditions.

We can also relate these cases to our more detailed case study of India. In India, government interest has evolved over time in relation to rural electrification, but comparable increases are not seen in the case of interest in clean cooking fuels. Institutional capacity has improved over time and, now, India compares favorably to most other countries in South Asia. Local accountability mechanisms remain mostly weak, with the exception of cases such as Kerala and West Bengal. While the country case studies are less detailed than that of India, it is notable that the same factors play a role. Even in Bangladesh, a country that faces many of the same challenges as India, the successes and failures of energy poverty eradication have reflected similar political-economic developments. These cases therefore testify to the external validity of the India analysis and provide a clear description of the political economy of energy poverty.

An interesting and, for policymakers, inspiring finding pertains to the role of new technology. In Kenya, a case of persistent policy failure so far, albeit encouraging developments lately, off-grid technologies have begun to fill the gap left by lack of success in grid extension. In Bangladesh, solar home systems have played an important role in alleviating energy poverty. While clean cookstoves have yet to achieve similar degrees of success outside the singular exception of China, their potential is high, provided challenges related to product affordability and quality can be solved.

On a final note, it is also important to consider the role of human agency in the case studies. While government interest is not something that can be easily manipulated, as donors trying to impose their preferences on reluctant governments have learned over the years, policy design can contribute to institutional capacity building and the strengthening of local accountability. Cases such as Bangladesh, where some parts of the energy sector are performing much better than others, are a stark illustration of this basic fact. Local accountability has often played a critical role in success stories through deliberate government efforts to strengthen it; this was true even for Mao's totalitarian regime in China.

7 Conclusion

Many consider energy poverty an administrative or technical problem, but this book has shown that success in improving energy access depends to a large extent on the government's political and economic incentives to act. While there is no denying that fundamentals such as energy resources, geography, and wealth are important covariates of energy poverty, an exclusive focus on them obscures as much as it illuminates. From Vietnam to Ghana and South Africa, a common thread runs through effective measures to combat energy poverty, and that thread is government interest. Where governments have strong political or economic reasons to combat energy poverty, there is an opportunity for effective policy. In turn, the extent of this success depends on the availability of institutional capacity in the energy sector and the presence of effective local accountability mechanisms.

The evidence for these claims comes from two sources. First, we have conducted a detailed case study of India's efforts to improve energy access. We showed that rural electrification levels and access to modern cooking fuel were both very low when the country became independent. While the country has made some progress in improving rural electrification, there has not been much progress in improving cooking fuel access, especially in rural areas. We have presented evidence that government interest for improving rural electrification is now at a high level and institutional capacity has been increasing since reforms in the 1990s. However, local accountability still remains low overall. Using five states with differing levels of government interest, institutional capacity and local accountability, we have shown that there is considerable spatial variation within the country. Some states were able to provide electrification to a large number of rural households, whereas others still lag behind.

The other qualitative approach consists of eleven comparative case studies across all regions and levels of wealth. We documented China's and Vietnam's stunning success in providing electricity to their populations, despite their extreme poverty at the time. We also showed that China was able to provide modern cookstoves to hundreds of millions of its citizens—a remarkable feat. South Africa clearly underscores the role of government incentives. Following the empowerment of millions of black South Africans in 1994, the government at once pursued vigorous policies designed to provide energy to its poorest people. Brazil and Chile shed light on two very different South American cases. Chile, despite being comparatively richer than our other cases, struggled to improve rural electrification rates, too. Only in the 1990s did it make decisive progress on that front.

The next five cases revealed more difficulty. Nigeria, despite abundant fossil fuel reserves, has made little progress in reducing energy poverty. The case of Bangladesh shows that government interest is insufficient on its own; institutional capacity and accountability are also necessary. Kenya, despite being a pioneer in off-grid solar systems, has struggled with grid extension. Finally, Senegal shows how a country made more progress in providing modern cooking energy than in expanding rural electricity access, as deforestation was a major problem for policymakers but the lack of rural electrification remained a minor nuisance.

This concluding chapter discusses these results and summarizes the most important implications of these findings. For research on energy poverty and access, we begin with a discussion of the role of political institutions. The evidence clearly shows that political incentives are central to any government effort to eradicate energy poverty, and future research should focus on developing precise measures of government interest. While standard considerations, such as regime type and the urban–rural power balance, are important, our case studies show that they are not sufficient.

Furthermore, there is considerable variation among interested governments in their ability to succeed. Scholars and practitioners have already recognized the importance of institutional capacity, but our findings on the modifying effect of political incentives are novel. There has also been far too little attention to the role of local accountability in both democratic and autocratic societies. To sum it all up, the most important challenge for future research on energy poverty is to develop strategies for identifying interested governments, measuring their local accountability and institutional capacity, and acting to improve them.

Beyond the study of energy poverty, our study has notable implications for political science and political economy more broadly. Energy poverty is but one facet of the broader challenge of dealing with abject poverty in the world. Political scientists have both correctly identified democracy as an important determinant of efficient pro-poor policy (Olson 1993; Lake and Baum 2001; Adserà, Boix, and Payne 2003; Brown and Mobarak 2008; Olken 2010) and identified the risk of dysfunctional democratic governance due to clientelism and patronage (Stokes 2005; Keefer 2007; Hagopian, Gervasoni, and Moraes 2009; Kohli 2009; Thachil 2011). However, these general determinants ultimately shape outcomes through concrete processes of policy formulation and implementation. In the comparison of rural electrification and provision of modern cooking fuels, both problems are solved through public service delivery, and yet government performance has varied widely. In particular, rural electrification is often of considerable interest to authoritarian regimes. Furthermore, institutional capacity is a sectoral, not cross-cutting, virtue, and local accountability manifests itself through concrete implementation support. Some recent studies have used electrification or lighting as a dependent variable (Hodler and Raschky 2014; Min 2015), but these studies miss some of the action by conceptualizing electricity as a public good or service in an abstract manner, without paying enough attention to the ground realities of rural electrification. We have sought to avoid this pitfall by adopting a bottom-up, contextualized approach to the subject matter, emphasizing the specific analytical challenges that energy poverty features.

Finally, we believe that our results are of interest to practitioners. Most important, searching for the holy grail of effective policy in a generic sense is a misguided effort. Government interest is the cornerstone of policy effectiveness, and future studies of policy design must begin with adequate evidence to show that the government exhibits genuine interest. Governments that lack interest may design impressive policies on paper to please donors and other audiences, but there is no reason to believe that they actually care about the quality of the outcome. Only among interested governments should questions such as institutional capacity, local accountability, and policy design be considered.

This simple observation is a call for new directions in the study of energy access policy. Energy access scholars should focus their efforts on analyzing the track record of policies implemented by enthusiastic governments instead of conducting additional case studies of policy failure in the case of governments that do not care about the outcome. In the absence of government interest, variation in institutional capacity and local accountability—perhaps, as well, other determinants of policy quality that we have not identified—cannot explain success and failure. Therefore, more policy-relevant knowledge can be inferred from studying policies and programs implemented by interested and committed governments. Furthermore, considerations of institutional capacity and local accountability should be at the forefront of these studies. By first accounting for these fundamental factors, scholars can then better isolate the effect of particular policy design on the outcome.

No treatment of energy poverty is complete without some reflection on sustainable development. Our most important contributions to this debate are threefold. First, our results show that due to considerable variation across national contexts, international rules that restrict the strategies available to the least developed countries can do considerable damage. While we recognize the importance of climate change mitigation for the future of the planet and human civilization, we also believe that the primary responsibility for addressing the problem lies with the wealthy global elites and middle class. An example of clashing priorities is the Overseas Private Investment Corporation's (OPIC) investment rules. OPIC is funded by the US government and private capital and channels funding to development projects in other countries. On paper, therefore, it ought to be at the forefront of the fight against energy poverty. Yet as a result of lawsuits filed by environmental NGOs and various US cities, OPIC was forced, starting in 2002, to reduce the greenhouse content of its portfolio (Moss, Pielke, and Bazilian 2014). In practice, the net result was that OPIC could not participate in grid extension projects that involve coal, oil, or even natural gas. Since 2009, OPIC has invested in a grand total of two fossil fuel power plants for a total capacity of about 350 megawatts—just one-third of a modern nuclear power plant (Moss, Pielke, and Bazilian 2014, 7).

This policy has created controversy among environmentalists and development specialists.[1] Aims such as economic and sustainable development do not always neatly align. In the short and medium runs at least, solving energy poverty may require the development of environmentally unfriendly resources. In such a perspective, the preferences of NGOs may sometimes clash. The rules imposed on OPIC were triggered by a lawsuit filed by Greenpeace; these were roundly criticized by other NGOs such as ONE, which "believe[s] narrow changes should be made to the policy that currently prevents OPIC from increasing investments in non-renewable technologies like natural gas that are needed to meet the electricity needs of the poorest and least carbon emitting countries in Africa."[2] Attempts to weaken the greenhouse emissions cap drew the ire of other NGOs and environmental activists. A number of NGOs now argue that the focus ought to be on clean energy first, pointing to cases such as Bangladesh in which small-scale, off-grid devices provide power to millions of people. Ultimately the best policy depends on the validity of assumptions about the scalability of these alternatives and their future cost, a complicated calculation. By showing the complexity of the energy issue, we have shed light on the nature of the trade-offs at stake.

Second, many of the fundamental challenges of basic electrification and modern cooking fuels lie within distribution. Basic household energy access does require energy generation capacity, but today's efficient technologies can remove abject energy poverty without massive additions to generation capacity. The energy required to address the basic needs of the world's energy-poor is insignificant compared to what the wealthy consume. Therefore, donors and international organizations worried about sustainability need not refrain from investing in grid extension efforts or in projects that replace traditional biomass with LPG.

Finally, we have also shown that many countries have made good use of new energy technologies, such as off-grid solar power. These technologies have now proven that they can play an important role in reducing energy poverty. They will not replace conventional power sources or centralized plants any time soon, but they create new opportunities for interested developing country governments to act. Instead of trying to stop conventional energy projects, environmental advocates should genuinely focus on securing additional funding for new projects that mitigate energy poverty while enabling the progress of clean technology. In this regard, efforts such as the United Nations' SE4ALL are welcome.

The Political Economy of Energy Poverty

Government policy is central to mitigating energy poverty, but not all governments have an interest in doing so. This observation is hardly revolutionary, but it has important implications for the social science of energy poverty. If we consider the mitigation of energy poverty first and foremost a policy challenge, then understanding the sources of government interest is the first stage of any comprehensive analysis of success and failure. Past studies of energy poverty have largely framed energy poverty as a problem of policy design and implementation (Sagar 2005; Ailawadi and Bhattacharyya 2006; Bhattacharyya 2006; Bazilian, Nakhooda, and Van de Graaf 2014), but this standard approach to the question is too narrow. Policy design and implementation become relevant only after a government commits itself to achieving success. We agree with Bazilian, Nakhooda, and Van de Graaf (2014), who emphasize the importance of governance and pro-poor policy, but we must add that good governance ultimately depends on the government's incentives to invest resources, time, and political capital.

As we have shown, government interest reflects both political and economic considerations. Both democratic and autocratic rulers can choose to invite the rural energy-poor to join their political support coalition. In democratic societies, this strategy reflects the electoral clout of the energy-poor; in autocratic societies, the risk of political instability and turmoil sometimes compels the government to act. Even governments that do not depend on the energy-poor for their political survival may choose to invest in mitigating energy poverty. Especially in the case of rural electrification, notable economic benefits are sometimes available. Perhaps the most important of these benefits is the use of electricity for groundwater extraction to boost agricultural productivity.

For scholars of energy poverty, the foundational role of government interest creates new research imperatives and opportunities. In an effort to reach beyond standard measures such as regime type, we have used a series of qualitative methods to shed light on the sources of government interest. We have found that governments of all stripes are savvy about the political and economic benefits of eradicating energy poverty, whether the ultimate goal is self-reliance, as in Mao's China, or electoral promises in the context of universal service delivery requirements, as in Lula's Brazil. Recognizing the limitations of our case studies, we believe it is essential for future studies of energy poverty to develop systematic, precise, and generalizable measures of such factors.

Understanding government interest is necessary for going beyond the foundations of energy resources, geography, and the level of economic development. We have seen that these factors are important. For example, Africa's low level of achievement in energy access largely reflects widespread poverty on the continent and the lack of resources for infrastructure investment. Analyses focusing on the fundamentals are not particularly interesting for policy, as their implications are obvious: wealthy countries have less energy poverty than poor countries do. Furthermore, the fundamentals are not sufficient for explaining variation in energy poverty across the world, as our case studies have shown. A key message of this book is that government interest plays a central role in determining the success and failure of energy poverty mitigation.

Besides the analytical and empirical challenges of measuring government interest, our theory and evidence challenge scholars of energy poverty to change their approach to institutional capacity. The importance of capacity is widely recognized in the literature (Mawhood and Gross 2014; Nygaard 2010; Reddy, Balachandra, and Nathan 2009; Rehman et al. 2012; Sovacool 2012b), but previous studies rarely consider the conditions that modify the importance of institutional capacity. Our findings suggest that institutional capacity is mostly relevant for governments that have political and economic incentives to eradicate energy poverty. Therefore, studies that evaluate the association between institutional capacity and energy poverty alleviation risk understating the true effect of institutional capacity.

To improve future research on energy poverty, scholars of institutional capacity should make a consistent effort to first verify that the country in focus has an interested government. Once government interest has been confirmed, it is time to begin considering institutional capacity. Our findings confirm that while state capacity in a generic sense is important, it is not the critical factor. Min (2015) reports that state capacity cannot explain rural electrification achievement, but we believe his results are somewhat misleading because they do not focus on capacity in the energy sector itself. Our suggestion for future research on energy poverty is to invest heavily in the improved measurement of sectoral capacity. In turn, this investment will enable theory development and systematic hypothesis testing.

While our primary concerns about the use of institutional capacity are the dependence of its relevance on government interest and the challenge of measurement, our concern about the role of local accountability is more fundamental. Although previous case studies and project evaluations have often noted that local participation and accountability to the rural energy-poor themselves are important for success (Barnes 2007; Bekker et al. 2008; Bhattacharyya and Ohiare 2012), there was no systematic theoretical or empirical treatment of the topic before our study. We have shown that in conditions characterized by strong government interest, local account-ability can significantly improve the transmission of information and the enforcement of competent implementation by officials in the context of energy access policy. In India, for example, increases in government inter-est and institutional capacity over time enabled progress in rural electrifi-cation, but the lack of effective local accountability mechanisms created difficulties. Even below the national level, states such as Gujarat performed surprisingly poorly because of the lack of strong accountability mecha-nisms in rural electrification. These strong results suggest that the analysis of local accountability must play a much more central role in future studies of energy access policy.

Here we recognize that we have barely scratched the surface. We have not tried to explain the origins of local accountability in the energy sector. What is the role of sectoral institutions and programs that enable citizen participation in regulatory and policy formulation processes? Answers to these questions are the logical next step for a full account of variation in the successful enactment and implementation of energy access policy.

Combining the fundamentals and political economy factors has the important advantage of laying the foundation for the inclusion of other factors, such as technological progress. An important result of our study is that new technology does not by any means reduce the importance of political economy. Even a largely private effort to use new technology, such as solar home systems or improved cookstoves, requires, at the very mini-mum, that the government enact and implement regulations that allow private business growth. Interested governments with high degrees of insti-tutional capacity and local accountability are in the position to capitalize on new technologies. New technology is important for eradicating energy poverty regardless of the constellation of political factors, but the exact mechanism through which technological progress maps into success in improving energy access depends on government interest, institutional capacity, and local accountability.

To summarize, we recommend the following approach to energy poverty for future scholarship. First, scholars need to reach beyond the fundamentals by recognizing, measuring, and analyzing government interest. Having done so, it is important to also adequately capture institutional capacity. Equally important but more challenging is the role of local accountability in energy poverty mitigation, the origins of which remain elusive to social science. Under this setup, social scientists can readily accommodate new factors, such as technological progress, into their heuristic and analytical models.

Broader Implications for Social Science

In writing this book, we have deliberately adopted a bottom-up approach to researching the subject matter. Instead of applying a generic theory of political economy to energy poverty, we have given priority to understanding energy poverty and tailored our theory according to this imperative. Nonetheless, our findings have some important implications for social science more broadly. While scholarly books published by social scientists typically begin with a foundation at the abstract level of concepts, we began with the specific problem of energy poverty and then applied the analytical tools provided by political economy to tackle the issue. Turning the standard approach on its head, we have noticed some important limitations of the top-down approach to hypothesis testing. While we do not advocate the disposal of the conventional hypothetico-deductive method, our experience and results suggest some important directions for the future of political economy.

One of the essential observations from the case studies in particular is the importance of properly accounting for elusive concepts such as government interest. In political economy, scholars have paid much attention to the effect of democratization on public goods provision (Lake and Baum 2001; Adserà, Boix, and Payne 2003; Keefer 2007; Brown and Mobarak 2008), because of the tremendous importance of the two topics and because they can be quantified in a precise manner. In the case of energy poverty,

Min (2015) has used the case of electricity access, measured as nighttime lighting, to test theories of democracy and public good provision. However, he himself notes that his book cannot establish whether democracy is good for the energy-poor: "To be clear, these propositions do not imply that democracy is necessarily better for the poor or that service provision is of higher quality in democratic settings" (9). A key reason for this hesitation is that his approach does not allow him to investigate the processes and implications of concrete energy access policies. By compromising on quantification and instead emphasizing the importance of solving the puzzle of variation in energy poverty, we have reached beyond provisional answers and made a more definitive statement: government interest, which is often driven by democratization, has clear benefits to the marginalized rural communities living in energy poverty. While some authoritarian regimes, such as China and Vietnam, have made impressive progress in reducing energy poverty, overall the evidence clearly shows that democratization promotes the welfare of the rural energy-poor.

The reason that political economists have sometimes expressed doubts about the autocracy-democracy difference is the persistence of dysfunctional policy in some democracies. This is one of the fundamental topics of study in modern comparative politics (Stokes 2005; Keefer 2007; Hagopian, Gervasoni, and Moraes 2009), and our results can speak to the general debate. We do not contend that factors such as clientelism or patronage do not cause inefficiencies in public good provision, but in the important case of energy poverty, the real problems are elsewhere. Electricity provision is useful to the rural energy-poor, and while clientelistic political behavior may result in inefficient electricity pricing, many of the great success stories of rural electrification, such as South Africa and Brazil, have reflected large public investments by governments interested in their electoral prospects. In these cases, public service delivery seems to have reflected clientelistic quid pro quo exchanges between politicians and voters, and yet the outcome has been beneficial to the rural energy-poor. The combination of strong government interest and institutional capacity, along with a modicum of local accountability, has enabled success without much evidence for the importance of programmatic political strategy.

Many democracies have, despite strong government interest, tripped over limited institutional capacity. In India, the fiscal problems of electric utilities have contributed to periods of slow rural electrification and, where the electric grid has appeared, frequent outages and voltage fluctuations due to poor technical performance. Arguably, these fiscal problems partly reflect the weakness of India's electoral democracy, and in this regard scholars who worry about dysfunctional electoral politics, such as Kohli (1987), have it right. And yet the story is much more complex than that. In some states, such as Kerala, inefficient electricity pricing has not raised any barriers to tremendous success in rural electrification. In other states, such as West Bengal, communist parties have been at the forefront of successful power sector reforms that have improved institutional capacity.

Local accountability is, of course, related to democracy in national politics. There is no denying that democracies often give their citizens much more authority in local issues as well. Yet our case studies have shown that authoritarian regimes, such as China and Vietnam, can achieve impressive degrees of local accountability. Moreover, some democracies, such as India, do a remarkably poor job of holding local authorities accountable for their actions. For the eradication of energy poverty, variation in local accountability has been an important factor in a number of cases. This reflects the inevitability of local implementation in the case of rural electrification and the provision of modern cooking fuels. Infrastructure is physical and cannot be constructed, maintained, or operated at arm's length.

We believe political economists have not put enough emphasis on local accountability in the analysis of public good provision and public service delivery. National incentives may allow effective policy with limited local accountability in some cases, but in most of the successful cases we have seen, local accountability has been an important channel of influence. The most successful cases of energy poverty alleviation have relied heavily on local officials, whose effective policies have been driven by the transmission of information across levels of government and by the direct link between performance and political survival. From rural infrastructure to service delivery and social policy, future studies in political economy have to pay more attention to the role of local accountability. It is insufficient to focus on performance at the village level, as Tsai (2007) has done in her important book on China's local democracy; future studies must also examine how local accountability modifies the effectiveness of national programs and policies. There is no reason to believe that the significance of our findings is limited to the case of energy poverty.

More generally, our analysis of two domains of energy poverty, electricity access and cooking fuels, and the role of new technology in the political-economic force field has shed light on the importance of considering the technical details of public service delivery. We have uncovered an important difference between rural electrification and modern cooking fuels that reflects, first and foremost, the different value of electricity and cooking fuels for productive uses. In an example of public service delivery by coincidence, many governments have made heavy investments in rural electrification, not because of the goal of pro-poor policy but for the simpler goal of boosting agricultural productivity. By doing so, these governments have alleviated energy poverty, most often in contexts featuring high levels of institutional capacity and local accountability.

The technical features of electricity have had fundamental political implications, and a misunderstanding of these technical features would have prevented us from considering the possibility of public service delivery by coincidence. Contemporary political science needs to pay more attention to such technical features and provide specialized training to scholars. Without such specialized training, political economy approaches risk being irrelevant and, at worst, creating more confusion than clarity because of fundamental errors in the analysis of the technopolitical interface.

To summarize, our approach to the study of energy poverty can be fruitfully applied to many problems of public good provision and public service delivery. A political economy approach meshes surprisingly well with a bottom-up approach that prioritizes the subject matter at hand, even if this requires departing from simplistic and generic theories that focus on abstract concepts, and pays close attention to technical details. For such topics, we recommend a combination of national and subnational case methods. Any study that gives priority to the subject matter will inevitably face the difficulty of considering a large number of contextual factors, as ignoring them will result in theory and evidence removed from the ground realities. Our experience suggests that to deal with this challenge, a combination of careful national and subnational evidence is very useful.

Policies to Mitigate Energy Poverty

What can a political economy approach say about policies to mitigate energy poverty? Most important, our findings highlight the importance

of beginning with the issue of government interest. While this observation may appear trite to some, the reality is that the vast majority of policy
analysis in the field of energy poverty steers clear of political questions. This
strategy might appeal to those who want to avoid conflicts and disputes,
but it carries great risks. There is not much that scholars or practitioners
can learn about the policy designs of governments that simply do not care
about the outcomes. It is encouraging to see that many governments now
show genuine interest in eradicating energy poverty. Yet examples such as
Nigeria show that the lack of government interest undermines policies and
investments that appear impressive on paper. This applies to donor projects
as well. Because of the fundamental importance of government interest,
donor projects in countries without government interest in success cannot
expect much.

Unfortunately, government interest is difficult to shape. Because it reflects
domestic political and economic considerations, it cannot be shaped by the
international policy community, except at the margin. For example, initiatives such as SE4ALL can spark government interest by providing examples
of the benefits of rural electrification, but if these benefits are not politically
or economically valuable to the government, there is little reason to believe
that awareness of them will shape policy formulation.

A more realistic approach to dealing with variation in government
interest is to identify governments that express it. In this regard, our study
validates the approach adopted by SE4ALL. In this program, developing
countries must express their interest and publicly commit to action to join
the initiative. Although it is unlikely that SE4ALL would have large effects
on government interest in the joining countries, the program can offer
participating governments access to useful resources and lessons that contribute to the building of institutional capacity and the creation of local
accountability mechanism. For example, international SE4ALL meetings
bring together government officials from participating countries, along
with business and civil society interests, for the purpose of information
exchange and peer learning.

At the time we are writing this paragraph, over seventy countries have
done so.[3] While some of these countries have an unimpressive history of
energy poverty mitigation—Nigeria is a SE4ALL member, after all—one can
only hope that these recent commitments reflect increasing interest. It is
certainly a promising sign that sub-Saharan African countries have shown

particular interest in SE4ALL, as Africa's recent economic success creates ample opportunities for productive public policy to reduce energy poverty. Among countries that show a high and stable or increasing level of government interest, the international policy community has a genuine opportunity in the development of institutional capacity. Our results show that one size does not fit all. There is no reason to debate the merits of different models, such as those emphasizing the leadership of rural electrification agencies versus offering incentives to electric utilities, in the generic sense. This kind of policy research provides little insight, given that our case studies show that variegated models can offer success in favorable political conditions. Instead, efforts to build institutional capacity should begin with the recognition of the genuine and legitimate political and economic constraints of host country governments. Social scientists have an important opportunity to contribute to energy poverty eradication through direct engagement with governments that have shown interest.

To be sure, strategies to build institutional capacity cannot be expected to produce results in the short run. In one case after another, we have seen that institutional capacity is not something countries can create when they need it. Instead, institutional capacity requires a sustained but adaptive sectoral strategy. We have seen that even the least developed countries, such as Bangladesh and Vietnam, can achieve impressive results in difficult conditions if they have the relevant sectoral capacity. In Bangladesh, for example, the top-down efforts were severely undermined by a lack of capacity, but rural electrification cooperatives and off-grid financing models achieved impressive success due to institutional capacity.

Encouragingly, these examples show that even in the short run, governments can tap into preexisting institutional capacity. The case of Bangladesh is particularly illustrative in this regard. The starkly differing degrees of success in top-down versus bottom-up electrification show that a country can possess considerable institutional capacity—and local accountability—in one domain while lacking it in another. The identification of technosocial strategies that rely on relevant institutional capacity offers an immediate opportunity for researchers and practitioners to increase the effectiveness of energy access policy.

While institutional capacity cannot be built quickly, our analysis suggests that local accountability offers more immediate opportunities. Because local accountability is both a general and a sectoral concern, national governments can reduce energy poverty by creating robust institutional mechanisms that allow the local population to express their views and reward officials for effective policy. While such mechanisms probably cannot work in the worst-governed local communities, cases from South Africa, China, Ghana, Chile, and Brazil show that local accountability also depends on national policy. For example, the requirement that local communities contribute to rural electrification efforts has proven effective by ensuring that the most enthusiastic communities receive priority for public investment and that local people have a direct interest in returns to public infrastructure investment.

We believe that the design of appropriate local accountability mechanisms is the next frontier for energy access policy. Of all the fundamental factors that we have emphasized, local accountability is the most malleable. Compared to government interest and institutional capacity, local accountability depends to a larger extent on policy design and the efforts of government officials at higher levels of administration. The approaches that appear successful in our case studies are community funding, local committees that monitor and evaluate projects, and, where possible, engagement with nongovernmental organizations that can reach out to the local community and facilitate financing.

If investments in local accountability promote energy poverty, they will ultimately create a positive reinforcement mechanism, as the eradication of energy poverty can itself contribute to political and economic development. As the famed economist Amartya Sen (1999) has argued, economic development is ultimately about expanding the freedoms and capabilities of the poor. These include both economic resources and political freedom, and energy poverty can contribute to the development of both, as the World Bank's former lead energy specialist Morgan Bazilian notes:

Access to energy enables each of those fundamental rights, which is why efforts to eradicate poverty cannot afford to ignore it. True, the barriers that impede progress on ending energy poverty are formidable: scarce financial resources, competing priorities, weak institutions, and the sometimes misguided interventions on the part of outsiders. But they are well within the world's ability to overcome, and they are far less imposing than many of the technical obstacles humanity has already vanquished. The laws of physics operate the same in South Asia and sub-Saharan Africa as they do in Europe and North America. The needed materials are simple things, such as steel, concrete, copper, and glass.[4]

Today, these issues are particularly important in sub-Saharan Africa, where energy poverty is more widespread than in any other region of the world. In its 2015 progress report for the continent, the Africa Progress Panel (2015, 15), chaired by Kofi Annan, emphasizes energy access and notes that

energy policy is at the heart of the opportunity. For too long, Africa's leaders have been content to oversee highly centralized energy systems designed to benefit the rich and bypass the poor. Power utilities have been centres of political patronage and corruption. The time has come to revamp Africa's creaking energy infrastructure, while riding the wave of low-carbon innovation that is transforming energy systems around the world. Africa cannot afford to stand on the sidelines of the renewable energy revolution. It can play its part in this revolution and tackle the challenges of transitioning away from fossil fuels.

While the panel appears to underestimate the importance of centralized solutions such as grid extension, its focus on energy as an opportunity is timely and welcome. Thanks to reduced rates of civil war and democratic consolidation, many African governments are, after several lost decades, in a position to enact and implement ambitious public policy to promote socioeconomic development. Our findings from countries such as South Africa suggest that this opportunity may result in rapid improvements in energy access, with positive effects on African economies more generally. Even in challenging environments such as Kenya, off-grid solutions are already making a difference.

Our findings also speak to the role of new technology in policy. As we have already noted, technological options should be tailored according to political, economic, and institutional capacity considerations. Consequently, debates on whether grid extension or off-grid renewables are *the* answer to energy access have little merit. For example, where large populations live under the grid but fail to secure household electricity connections due to high charges (Lee et al. 2014), there are technical opportunities for both grid electrification and off-grid development. The relative importance of each strategy depends not only on technical considerations, such as the market prices of relevant technology, but also on institutional capabilities, the regulatory framework, and political resistance to change. In Kenya, where Lee et al. (2014) provide their detailed evidence on large numbers of unelectrified households living in the immediate vicinity of the grid, our case study has shown that a plausible explanation for this phenomenon is the lack of institutional capacity of the power sector administration, rural electrification agencies, and electric utilities. In such a case, the mix of technological approaches should depend on an assessment of why there are households living under the grid.

We do need to caution against those renewable energy enthusiasts who believe that grid extension has been a failure. For example, consider Jigar Shah, the founder of the major solar company SunEdison, who writes the following about distributed renewable energy:

> The truth is that an over-reliance on centralized grid extension and large-scale power plants will keep a billion people in the dark. It is time to recognize what even the IEA says is overwhelmingly necessary, but dramatically under-invested in: distributed renewable energy for those living beyond the grid.[5]

We agree that distributed energy has a lot of potential and will play an increasingly important role in the future, but the historical record is full of remarkable success stories in eradicating energy poverty through grid extension. Based on the available evidence, investments in grid extension can make major contributions to the eradication of energy poverty, provided interested governments have the institutional capacity and local accountability mechanisms required for efficacious action. Besides, there is no reason that a developing country's electricity generation mix for the national grid cannot contain a substantial share of modern renewables.

In the case of cooking, we see similar issues. While success in the provision of modern cooking fuels is less common than success in rural electrification, we have shown that the lack of government interest is a key reason. If governments show more interest in the future, perhaps as a result of increased awareness about the economic value of improved public health, there is no reason that the use of modern cooking fuels cannot grow rapidly even in the least developed countries. With the exception of China's success in efficient cookstoves and South Africa's electric cooking, however, it is notable that most successes so far rely on the distribution of LPG stoves and fuel. Perhaps most remarkable of all the cases we have considered is Senegal, where rural electrification is lacking but LPG has become a major cooking fuel.

For efforts such as SE4ALL, these results highlight the importance of technological neutrality. The international policy community must escape the trap of tribal allegiance to a specific technology. New technology offers great opportunities for improved policy, but the exact nature of these opportunities depends on the local context. Debates about the relative importance of grid extension versus off-grid electrification, or LPG versus improved biomass cookstoves, are futile, distracting, and confusing. The world needs less such debate and more emphasis on concrete, sustained, and customized action in specific contexts.

Sustainable Energy Access in the Anthropocene?

Besides the debate on appropriate national policy, energy access provokes heated exchanges because of its potential environmental implications. The use of fossil fuels for energy is the single most important cause of climate change. As a result of growing energy consumption in the developing world, scientists expect humanity's carbon dioxide emissions to grow rapidly in the coming years. Some economists suggest that today's energy trajectory models underestimate the future growth of energy demand for energy by a large margin (Wolfram, Shelef, and Gertler 2012). If efforts to improve energy access result in the increased use of fossil fuels, they could exacerbate global warming, with severe effects on nature and society.

The environmental aspects of energy access are part of the broader challenge of human-environment interactions in what Crutzen (2006) calls the Anthropocene—the era of human dominance of our natural ecosystems. Due to growing wealth and population, our species now dominates the planet's ecosystems. The single most important force shaping the natural environment is human society. Improved energy access magnifies the human effect on the environment by increasing the pressure on the planet to fuel social and economic activity. As energy access improves, poor people who previously lived without electricity and modern cooking fuels will use appliances such as televisions and refrigerators. If powered by fossil fuels, such appliances cause environmental destruction across the globe.

To be sure, energy access is but one piece of the climate change puzzle. As Pachauri (2014) shows, increased household energy access in India—a country wealthier than most of sub-Saharan Africa—would cause a "trivial" addition to global greenhouse gases. The reason is that because most Indian households, still living in poverty, use very little energy, each additional electricity connection would not generate much carbon dioxide emissions. The same would probably be even more true of poorer countries, where energy access is an even larger problem. Moreover, an Overseas Development

Institute (2016) report on going "beyond coal" in energy access policy correctly notes that in the case of rural electricity access, the problem is not the generation but the distribution of electricity (ODI 2016). But this does not mean that we can safely ignore energy access as an environmental issue in the long run. Over time, households begin to use more and more electric appliances and other devices that require modern energy. Consequently, the long-term effects of access to modern energy are probably much larger than what Pachauri (2014) estimates based on historical data.

There is no denying that energy access is a fundamentally important development goal. Lacking access to energy is a stringent constraint on the ability of the world's poor to realize their aspirations and potential. In the absence of modern energy, life is hard, and the global environment is hardly a relevant concern. Therefore, policies that restrict energy access to mitigate climate change carry a high cost for the world's poor, while citizens of industrialized countries and the global middle class enjoy convenient lives powered by fossil fuels.

Of course, only a true extremist would propose that the international development community deliberately prevent energy access. Instead, the challenges are more subtle. Should international development agencies and bilateral donors support fossil fuel projects? Or should they fund only renewable energy? Should these agencies and donors fund rural electrification through grid extension in countries that rely heavily on coal for their electricity? Or should they insist on aggressive policies to promote decarbonization? If coal is considered too dirty for investment, what about natural gas and large hydroelectric dams?

Our findings suggest that restricting energy access policy is dangerous because it can perpetuate and exacerbate energy poverty. Our case studies show that countries have relied on a variety of possible technological solutions to combat energy poverty. Some, such as South Africa, have had great success through grid extension. Others, such as China, began with distributed energy and later connected these facilities with the national grid. Yet others, such as Bangladesh, provide basic electricity access through solar panels. In Senegal and Brazil, LPG is now frequently used, whereas China chose to focus on efficient cookstoves and achieved phenomenal success with this strategy.

International policies that prevent countries from choosing their solution could undermine such success stories in the future. Perhaps more worrying, the countries that need such success the most are the ones that are the most dependent on international assistance. Large emerging economies such as India or Indonesia can mitigate energy poverty without external assistance, but the poorest of the poor, such as Liberia and Eritrea, cannot. Environmental rules set at the international level will make the world's wretched pay the highest price for an environmental problem that they have not caused, and this outcome would be an affront to justice. Emerging economies would steam ahead and achieve higher and higher standards of living, producing negative environmental effects, while the poorest of the poor would struggle. This is neither a solution to our global environmental problems nor fair or equitable.

Even in relatively poor countries, today's global investment realities may prevent the use of stringent environmental safeguards. Consider, for example, the 1,870 megawatt Gibe 3 Dam in Ethiopia.[6] Initially supported by the World Bank and the African Development Bank, the project was dropped after an environmental impact assessment commissioned by the Ethiopian government was released. The construction of the dam was completed by 2016, however, with funding from the state-owned Industrial and Commercial Bank of China.[7] Regardless of the net merits of the project, it is notable that losing the support of two major development agencies did not stop the project because Chinese capital was available for it.

Advocates of low-carbon energy sources may counter our argument by saying that there is no longer any need for fossil fuels. This may or may not be true, but if it is, then there is no need for any international restrictions on the use of fossil fuels either. If energy sources such as solar and wind are as competitive as their advocates claim, they will become the default choice for developing countries. As a result of such "leapfrogging" (Soete 1985; Zerriffi and Wilson 2010), economic development would no longer have to cause environmental destruction as much as it used to be. Therefore, we believe that renewable energy enthusiasts should not disagree with us. Only people who believe renewable energy to be very expensive but necessary for environmental reasons have valid reasons to contest our proposal, and even they have to acknowledge that their proposal will increase the cost of energy in very poor countries.

Another reason for rejecting restrictions on energy access is that they may poison climate negotiations. As the importance of the global South increases, any policies imposed by industrialized countries will make negotiations even more difficult than they already are. Restrictions on energy access projects in the very poor countries could turn climate change policy into an illegitimate cause, especially if these restrictions become a symbol of North–South conflicts. Scholars of North–South dynamics in international environmental negotiations have shown that concerns about equity are a major obstacle to progress (Sell 1996; Najam, Huq, and Sokona 2003; Najam 2005), and whatever marginal environmental benefits such restrictions bring about would be far outweighed by the worsening gridlock at the international arena.

Far from suggesting a passive strategy, all this points to the importance of new environment development coalitions and alliances. Industrialized and emerging economies have a genuine opportunity in framing the climate change challenge as a positive one. After the failure of the 2009 Copenhagen climate summit to produce a legally binding treaty, a group of scholars came together and wrote a document, the "Hartwell Paper," to propose alternatives to multilateral environmental diplomacy (Prins et al. 2010). One of their suggestions was to make energy poverty a more prominent theme in climate negotiations through what international scholars call an "issue linkage" (Stein 1980). By making a strong international commitment to eradicating energy poverty, the wealthier countries could make the entire climate change agenda more palatable and interesting to the world's poorer countries and their governments.

More generally, a positive approach to energy poverty alleviation creates an opportunity to showcase the benefits of clean technology. We have seen solar power play an important role in eradicating energy poverty. Such success stories can play an important role in enhancing the public legitimacy of renewable energy, a key issue in the spread of the technology, as Wüstenhagen, Wolsink, and Bürer (2007) have argued. If technologies such as renewable energy prove their value in combating energy poverty, this achievement may create political support for renewables among developing countries with legitimate concerns about the effect of climate mitigation measures on their economic development. For example, the government of a country that uses renewables to make progress in the battle against energy poverty, such as Bangladesh today or perhaps Kenya in the future, can hardly sustain the view that renewable energy is at best a distraction from more important energy issues and at worst a danger to economic development.

To finish this book on a positive note, we emphasize the opportunities that lie ahead. Many governments across the world have correctly identified energy poverty as a key socioeconomic problem, and the combination of government interest, institutional capacity, and local accountability mechanisms has allowed them to overcome limits set by geography, energy resources, and lack of wealth. New energy technologies have played an increasingly important role in this process during the past decade alone. While progress has been much faster in rural electrification than in the provision of modern cooking fuels, the case for addressing the negative environmental and public health effects of traditional biomass is strong. As developing countries industrialize and the importance of human capital grows in the global economy, the prospects for repeating the success stories of rural electrification in the provision of modern cooking fuels are brighter than in the past.

Notes

Chapter 1

1. Available at https://data.worldbank.org/indicator/EG.USE.PCAP.KG.OE (accessed November 20, 2017).

2. "Secretary Clinton Announces the Global Alliance for Clean Cookstoves at the Clinton Global Initiative," Media Note, Office of the Spokesman, Washington, DC, September 21, 2010, https://2009-2017.state.gov/r/pa/prs/ps/2010/09/147488.htm (accessed April 2, 2018).

3. See http://www.un.org/sustainabledevelopment/sustainable-development-goals/ (accessed December 1, 2017).

4. We consider China, Vietnam, South Africa, Ghana, Brazil, and Chile as relatively successful cases of energy access policy; Bangladesh, Indonesia, Nigeria, and Kenya were much less successful. Senegal is an unusual case in that it has seen plenty of success in improving access to modern cooking fuels, with much less progress in rural electrification.

Chapter 2

1. As we explain in this chapter, in practice organizations such as the IEA often rely on shortcuts, such as access to a minimal number of kilowatts per year, for measuring energy access. However, these thresholds are not definitional.

2. Of course, success is not guaranteed. In some countries, even urban regions have blackouts. The elites may not care about urban dwellers, and slums are often energy poor (Aklin et al. 2015).

3. However, accounts differ on the role and importance of coal (Clark and Jacks 2007).

4. For example, the 2001 energy crisis in Brazil, which is largely dependent on hydropower, was caused by a major drought.

5. For the United States, see "How Much Electricity Is Lost in Transmission and Distribution in the United States?" Energy Information Administration, https://www.eia.gov/tools/faqs/faq.cfm?id=105&t=3 (accessed December 10, 2017). For other countries, see World Development Indicators at https://data.worldbank.org/indicator/EG.ELC.LOSS.ZS (accessed December 10, 2017).

6. "Ethiopia: Dealing with the 'Last Mile' Paradox in Rural Electrification," Global Partnership on Output-Based Aid, http://www.gpoba.org/node/345 (accessed January 22, 2015).

7. Although nonsolid fuels are the modern and durable solution to access to cooking fuels, we acknowledge that there is a difference between the use of traditional biomass and resources such as charcoal. Because charcoal has a higher energy content per unit than firewood, it is an improvement over the latter in terms of convenience and benefits. However, charcoal production can contribute to deforestation and therefore is unlikely to provide a sustainable permanent solution to cooking fuel poverty.

8. As Douglas F. Barnes pointed out to us, one major difficulty is that widespread cooking with electricity increases peak power demand because households tend to prepare and eat dinner.

9. For instance, see "ANOMENA Improved LPG Cookstove," Clean Cooking Catalog, http://catalog.cleancookstoves.org/#/stoves (accessed January 28, 2015).

10. This database is publicly available at http://www.iea.org/energyaccess/database/ (accessed December 11, 2017).

11. Chile (100 percent), Mexico (99 percent), and Turkey (100 percent) are clearly OECD member states, but we nonetheless added them to show their electrification status. Because IEA (2017) does not provide any data for these three countries, we instead use electrification rates from the SE4ALL Global Tracking Framework (SE4ALL 2017a, data annex 1).

12. As with the map for electrification rates, we report access to clean cooking fuels for Chile (97 percent) and Mexico (86 percent) despite their OECD membership, while data for Turkey is not available. We also add data for the United Arab Emirates (100 percent) and Saudi Arabia (100 percent) from the SE4ALL Global Tracking Framework (SE4ALL 2017a, data annex 1). For Libya (99 percent), we use IEA (2017) data from 2010, the latest available.

13. These data are available from http://www.reegle.info/countries/paraguay-energy-profile/PY (accessed January, 30 2015).

14. Interview with a Brazilian electricity policy expert in Rio de Janeiro, August 2014.

15. "A Brightening Continent," *Economist*, January 17, 2015, https://www.economist.com/news/special-report/21639018-solar-giving-hundreds-millions-africans-access-electricity-first (accessed February 3, 2015).

16. These capacity constraints often stem from policies that prevent utilities from recovering their costs, thus making investment unprofitable.

17. "Dirty Pretty Rock," *Foreign Policy*, January 29, 2015, http://foreignpolicy.com/2015/01/29/dirty-pretty-rock-peak-coal-energy/ (accessed February 3, 2015).

18. "Our High-Energy Planet: A Climate Pragmatism Project," Breakthrough Institute, April 8, 2014, https://thebreakthrough.org/index.php/programs/energy-and-climate/our-high-energy-planet (accessed February 3, 2015). For the report itself, see BTI (2014).

19. "Poor Countries Shouldn't Sacrifice Growth to Fight Climate Change," Bloomberg Business, August 29, 2014, http://www.bloomberg.com/bw/articles/2014-08-29/poor-countries-shouldn-t-sacrifice-growth-to-fight-climate-change (accessed February 3, 2017).

20. "Two Videos That Illuminate Energy Poverty," *Gates Notes*, June 25, 2014, https://www.gatesnotes.com/Energy/Two-Videos-Illuminate-Energy-Poverty-Bjorn-Lomborg (accessed February 3, 2015).

21. "Gujarat to Be First State to Ink MoU with UN Energy Initiative SE4ALL," *Business Standard*, January 9, 2015, http://www.business-standard.com/article/economy-policy/gujarat-to-be-first-state-to-ink-mou-with-un-energy-initiative-se4all-115010900746_1.html (accessed February 3, 2015).

22. "About IRENA," http://www.irena.org/about/About-IRENA/aboutirena (accessed February 3, 2015).

23. "Renewables, Not Coal, Way Out of Energy Poverty in Africa," *CleanTechnica*, November 13, 2014, https://cleantechnica.com/2014/11/13/renewables-coal-way-energy-poverty-africa/ (accessed February 3, 2015).

24. The relevance of this question to energy poverty eradication depends on how minimal a level of energy access is considered enough for a household to escape energy poverty.

Chapter 3

1. Because our focus is on the eradication of energy poverty, we assume throughout that access to modern energy is technically feasible and not prohibitively costly. A different analytical approach is necessary for understanding the early years of electrification in the late nineteenth and early twentieth centuries (Hughes 1993; Kale 2014b; Shamir 2013).

2. "IMF Is Watching Former Darling Ghana Closely," *IOL Business Report*, April 10, 2015, https://www.iol.co.za/business/opinion/imf-is-watching-former-darling -ghana-closely-1.1842999 (accessed April 10, 2015).

3. In contrast, we do not see a serious time-inconsistency problem in the initiation of energy access programs. The first step in such programs is typically the extension of infrastructure to communities and households. If the financial resources are available, such extension can be initiated quickly to bring visible benefits to citizens. If a government wants to capture votes with energy access policy, it can do so even if the elections are only a few years ahead.

4. "Le Programme d'électrification rurale global (PERG) au Maroc," French Development Agency (Agence française du développement), https://www.afd.fr/home/AFD/ developpement-durable/DD-et-strategies/rioplus20/projets-rio20/electrification -maroc (accessed February 10, 2015).

5. The conversion to 2014 dollars is based on the US CPI. "Loan Agreement (Village Electrification Project), Loan Number 1695 MOR," http://documents.worldbank .org/external/default/WDSContentServer/WDSP/MNA/2013/09/05/090224b081e 8e49b/1_0/Rendered/PDF/Morocco000Vill0greement000Conformed.pdf (accessed February 10, 2015).

6. World Bank, "Report No. P-5261-MOR," August 30, 1990, http://documents .worldbank.org/external/default/WDSContentServer/WDSP/IB/1990/08/30/000009 265_3960928223502/Rendered/PDF/multi_page.pdf (accessed February 10, 2015).

7. Many democratic countries also have electoral institutions that overallocate legislative seats to rural areas in what political scientists call "malapportionment" (Samuels and Snyder 2001; Bayer and Urpelainen 2016).

8. Note, however, the difference between local accountability as a mechanism of information provision and the use of improved information to enhance accountability. As Lieberman, Posner, and Tsai (2014, 77) note, new information contributes to local accountability only when citizens gain access to new, relevant information that makes them mobilize to demand better services from government officials.

9. "480,000 New Solar Home Systems for Bangladesh," *CleanTechnica*, July 14, 2014, https://cleantechnica.com/2014/07/14/480000-new-solar-home-systems -bangladesh/ (accessed February 15, 2015).

10. Nancy Wimmer, "The Next Million: Grameen Shakti—An Innovative Model for Rural Energy Business," *Next Billion*, December 17, 2012, https://nextbillion.net/ the-next-million/ (accessed March 28, 2018); and Hans-Josef Fell, "The German Experience on the Way to 100% Renewabels," July 12, 2017, https://mm.boell.org/ sites/default/files/uploads/2017/07/myanmar_part_1_engl_july_2017_fell.pdf (accessed May 1, 2018).

11. "Clean Cookstoves Designed in Bangalore Are Transforming Lives," *Times of India*, September 2, 2014.

12. See http://cleancookstoves.org/partners/item/22/669 (accessed April 24, 2015).

13. A possible adverse consequence of this dynamic for the population is that the government uses basic electricity access through off-grid solar power as an excuse not to extend the grid.

14. To be sure, it is not clear that off-grid electrification efforts can contribute to economic productivity as much as grid extension does. Since productive loads of energy require a lot of electricity generation capacity, the total cost of providing enough electricity without a grid connection may be too high to be profitable. So far, distributed energy generation has, despite some efforts in various countries (Zerriffi 2011), yet to provide a substitute for grid extension in regard to the productive uses of electricity.

15. See Batley and Mcloughlin (2015, 281) for an analysis of how visibility and other characteristics of public services shape government incentives.

16. The numbers are based on approved "Energy Transmission and Distribution" operations for 2017, available from the World Bank's "Projects & Operations." See http://projects.worldbank.org/search?lang=en&searchTerm=&sectorcode_exact=LT (accessed December 10, 2017).

17. These numbers were obtained by AidData's database with searches for "rural electrification" and "cooking" for all donor countries and all possible recipients over all possible years. See http://aiddata.org/dashboard#/advanced/project-list (accessed December 10, 2017).

Chapter 4

1. This capacity is so meager that it is approximately the capacity of a single unit of one nuclear reactor in the United States, a country with more than one hundred commercial reactors today.

2. The socioeconomic survey of the National Sample Survey Office is typically run every six or seven years and covers questions about education, employment, expenditure, health care, and other topics. In the sixty-sixth round conducted in 2010, respondents were asked about both their primary source of energy for lighting and electricity expenditures. Here, we define electricity use as any nonzero expenditure on electricity. This definition includes both grid connections and other sources, such as paying a monthly fee to the operator of a diesel generator. If we had used the question about primary source of energy for lighting instead, the percentage would have been lower. See NSS (2012b) for details.

3. This number is calculated based on 27 percent of the rural population of India in 2011 (Government of India 2011b).

4. We discuss five main Indian states in this chapter. Orissa is located in eastern India and is a relatively poor state compared to other parts of the country. Uttar Pradesh is located in northern India and is the most populous but also one of the poorest states in the country. West Bengal is located in eastern India and is a relatively wealthy state compared to other parts of the country. Gujarat is located in western India and is among the wealthiest states in the country. Maharashtra is located in western India and is the wealthiest state in the country.

5. We discuss the implications of these colonial investments in more detail below.

6. As Kale (2014b) shows, urban electrification was initiated in the colonial era.

7. The plateau and drop in the number of villages during the late 1990s stem from a change in the definition of a village.

8. These numbers were calculated using two rounds of the India Human Development Survey (IHDS) conducted in 2007 and 2012 (Desai et al. 2007; Desai, Dubey, and Vanneman 2012). The specific variable used was the number of hours of electricity access per day for rural households.

9. As Kale and Mazaheri (2014) note, the rural, lower-caste social mobilization that has swept across Uttar Pradesh and Bihar has been conspicuously absent from Orissa.

10. These numbers were calculated using two rounds of the India Human Development Survey conducted in 2007 and 2012 (Desai et al. 2007; Desai, Dubey, and Vanneman 2012). The specific variable used was the number of hours of electricity access per day for rural households.

11. Computed using the Indian rupee–US dollar exchange rate at the end of 1990 from https://www.rbi.org.in/scripts/PublicationsView.aspx?id=15268 (accessed February 1, 2017).

12. According to Narayan (2009), higher-caste political activists have also successfully courted lower-caste voters by emphasizing communal conflicts between Hindus and Muslims.

13. These numbers were calculated using the second round of the India Human Development Survey conducted in 2012 (Desai, Dubey, and Vanneman 2012). The specific variable used was the number of hours of electricity access per day for rural households.

Chapter 5

1. "Lake Volta," *Encyclopaedia Britannica*, https://www.britannica.com/EBchecked/topic/632445/Lake-Volta (accessed April 10, 2015).

2. "Who We Are," Public Utilities Regulatory Commission, http://www.purc.com. gh/purc/features/fusion-menu (accessed July 20, 2016).

3. "PURC Orders ECG to Suspend Billing Software," *GhanaWeb*, May 24, 2016, https://www.ghanaweb.com/GhanaHomePage/NewsArchive/PURC-orders-ECG-to -suspend-billing-software-441532 (accessed July 20, 2016).

4. "IMF Is Watching Former Darling Ghana Closely," *IOL Business Report*, April 10, 2015, https://www.iol.co.za/business/opinion/imf-is-watching-former-darling -ghana-closely-1.1842999 (accessed April 10, 2015).

5. "Energy Crisis in Brazil Is Bringing Dimmer Lights and Altered Lives," *New York Times,* June 6, 2011 (accessed April 10, 2015).

6. See http://www.reegle.info/countries/brazil-energy-profile/BR (accessed April 10, 2015).

7. See http://www.reegle.info/countries/brazil-energy-profile/BR (accessed April 10, 2015).

8. See https://www.iea.org/policiesandmeasures/pams/brazil/name-24303-en.php (accessed April 2, 2018).

9. See https://blogs.worldbank.org/growth/brazil-announces-phase-two-growth -acceleration-program (accessed October 30, 2017).

10. See https://data.worldbank.org/indicator/SP.URB.TOTL.IN.ZS (accessed December 4, 2017).

11. See http://www.ultra.com.br/Show.aspx?IdMateria=BshizMPs7HOIdYjt0gITvw ==&linguagem=en (accessed October 30, 2017).

12. "An Unexpected Setback," *Economist*, June 1, 2012.

Chapter 6

1. This is perhaps unsurprising, given that USAID was one of Bangladesh's policies most important supporters and funders, but the virtues of the Bangladeshi policies are widely acknowledged (Taniguchi and Kaneko, 2009).

2. "Hasina Arrested, Sent to Sub-Jail," *Daily Star*, July 17, 2007, http://archive .thedailystar.net/2007/07/17/d7071701011.htm (accessed March 27, 2015).

3. "Electricity Tariff," Thakurgaon Palli Bidyut Samity, http://www.pbs.thakurgaon. gov.bd/ (accessed April 30, 2018).

4. "Bangladesh Installed 3 Million+ New Residential Solar Systems," *CleanTechnica*, November 19, 2014, https://cleantechnica.com/2014/11/19/bangladesh-installed-3 -million-new-residential-solar-systems-since-may/ (accessed April 20, 2015).

5. "Bangladesh Installs 3 Million Residential Solar Systems," *PV Magazine*, November 11, 2014, https://www.pv-magazine.com/news/details/beitrag/bangladesh-installs-3-million-residential-solar-systems_100017123/ (accessed May 30, 2015).

6. "Bangladesh's Constitution of 1972," Constitute, https://www.constituteproject.org/constitution/Bangladesh_2011.pdf (accessed October 30, 2016).

7. Rural Electrification Board Ordinance, 1977, http://bdlaws.minlaw.gov.bd/print_sections_all.php?id=557 (accessed March 27, 2015).

8. Interestingly, the push for private involvement has had effects on cooking, too. Data are scarcer, but it is estimated that Grameen Shakti alone had installed over 130,000 modern cooking stoves by the year 2010 or so (Sovacool and Drupady 2011). Unsurprisingly, these numbers lag behind electricity, but they attest to the power of new technologies in shifting the field.

9. "Indonesian President Joko Widodo Pledges to Cut Investment Barriers," *Wall Street Journal*, December 7, 2014, https://www.wsj.com/articles/indonesian-leader-joko-widodo-pledges-to-cut-investment-barriers-1417988251 (accessed December 10, 2017).

10. All data come from the World Bank's World Development Indicators, https://data.worldbank.org/products/wdi (accessed March 23, 2015).

11. After its independence from the United Kingdom in 1960, years of civil war (1967–1970), and military rule for almost thirty years, the 1999 presidential elections are often recognized as the first relatively free elections, marking the country's democratic renewal.

12. "Nigeria to Boost 2016 Budget Even as Crude Revenue Declines," Bloomberg, December 8, 2015, https://www.bloomberg.com/news/articles/2015-12-08/nigeria-boosts-2016-budget-by-a-fifth-even-as-oil-revenue-slumps (accessed November 14, 2016).

13. "Nigeria's NNPC 'Failed to Pay' $16bn in Oil Revenues," *BBC News*, March 15, 2016, http://www.bbc.co.uk/news/world-africa-35810599 (accessed March 15, 2016).

14. "Nigerian Oil-Revenue Audit Published," *Wall Street Journal*, April 27, 2015, https://www.wsj.com/articles/nigerian-oil-revenue-audit-report-published-1430164157 (accessed November 14, 206).

15. "Nigeria's Oil Company Kept Nearly Half of Oil Revenue," *Reuters*, June 29, 2015, https://www.reuters.com/article/nigeria-oil-idUSL5N0ZF4MC20150629 (accessed November 14, 2016).

16. "Nigeria's NNPC 'Failed to Pay' $16bn in Oil Revenues."

17. "Olusegun Obasanjo: 'We Must Mobilise Investment in to Africa," *AB2020*, May 16, 2016, http://africabusiness2020.com/2016/05/16/olusegun-obasanjo-its-time-to -tell-the-true-story-of-africa/ (accessed November 7, 2016).

18. "How Nigeria's Presidential Elections Works," *BBC News*, March 27, 2015, http://www.bbc.co.uk/news/world-africa-31111572 (accessed November 7, 2016).

19. "Privileged Nigerians Shouldn't Downplay Poverty Just Because It Makes Us Look Bad," *Guardian*, August 11, 2016, https://www.theguardian.com/commentisfree/2016/ aug/11/privileged-nigerians-downplay-poverty-western-media (accessed November 7, 2016).

20. "Nigeria No Longer Africa's Top Oil Producer, As Militants Cut Production," *Newsweek*, May 17, 2016, http://newsweek.com/nigeria-no-longer-africas-top-oil -producer-militants-cut-production-460676 (accessed November 14, 2016).

21. "Study Breaks Down Nigerian Solar Power Failure," *SciDevNet*, January 19, 2016, https://www.scidev.net/global/energy/news/nigerian-solar-power-failure.html (accessed July 12, 2016).

22. "Why Electricity, or a Lack of It, Is an Election Concern for Many Nigerians," *VICE News*, March 27, 2015, https://news.vice.com/article/why-electricity-or-a-lack -of-it-is-an-election-concern-for-many-nigerians (accessed April 1, 2015).

23. "Why Electricity, or a Lack of It, Is an Election Concern for Many Nigerians."

24. "In Nigeria's Election, Muhammadu Bihari Defeats Goodluck Jonathan," *New York Times*. March 31, 2015, https://www.nytimes.com/2015/04/01/world/africa/ nigeria-election-muhammadu-buhari-goodluck-jonathan.html (accessed April 1, 2015).

25. The Global Alliance for Clean Cookstove database is available online at http:// carbonfinanceforcookstoves.org/tools/projects/ (accessed April 1, 2015).

26. For further information on the Renewable Energy Programme, see http:// renewableenergy.gov.ng/projects/ (accessed April 1, 2015).

27. Sources diverge, but even more optimistic observers believe that rural electrification rate is about 4 percent (Kapika and Eberhard 2013, 44).

28. Press release by Robert F. Godec, US ambassador to Kenya, February 4, 2014, https://ke.usembassy.gov/u-s-ambassador-to-kenya-robert-f-godec-remarks-for-the -national-clean-cookstoves-and-fuel-conference-in-kenya (accessed April 6, 2015).

29. For a summary, see the Global Alliance for Clean Cookstoves website: http:// carbonfinanceforcookstoves.org/implementation/case-studies/the-efficient-cook -stove-programme-in-kenya/ (accessed April 6, 2015).

30. "Kenya's Energy Revolution: Full Steam Ahead for Geothermal Power," *Guardian*, November 22, 2013.

31. "L'ex-boss de l'ASER condamné à 5 ans ferme," *SenePlus*, July 18, 2013, (accessed April 28, 2015).

32. "Toujours aussi virulent!" *GFM*, August 13, 2014, https://www.gfm.sn/toujours -aussi-virulent/ (accessed April 28, 2015).

33. "Sénégal: Wade et le péril jeune," *Jeune Afrique*, August 3, 3011, http://www .jeuneafrique.com/190707/politique/s-n-gal-wade-et-le-p-ril-jeune/ (accessed April 28, 2015).

Chapter 7

1. "Unleash OPIC: Why Are We Fighting Climate Change on the Back of the World's Poor?" Todd Moss (Center for Global Development), June 20, 2012; "Electrifying Africa–But at What Cost to Africans?" *Foreign Policy in Focus*, September 16, 2013, http://fpif.org/electrifying-africa-cost-africans/ (accessed July 23, 2015).

2. "Five Questions about Electrify Africa and the Environment," *ONE*, September 13, 2013, https://www.one.org/us/press/5-questions-about-electrify-africa-and-the -environment/ (accessed July 23, 2015).

3. See http://www.se4all.org/actions-commitments/country-level-actions/ (accessed December 7, 2017).

4. Morgan Bazilian, "Power to the Poor: Provide Energy to Fight Poverty," *Foreign Affairs*, March–April 2015, https://www.foreignaffairs.com/articles/143063/morgan -d-bazilian/power-to-the-poor (accessed February 25, 2015).

5. "Sorry Bill Gates, You Are Wrong on Renewable Energy." *Greentech Media*, August 22, 2014, https://www.greentechmedia.com/articles/read/sorry-bill-gates-you-are -wrong-on-clean-energy (accessed February 22, 2104).

6. "China's Backing for Ethiopia Dam Riles Activists," June 1, 2011, *Mail and Guardian Africa*, https://www.cbsnews.com/news/chinas-backing-for-ethiopia-dam-riles -activists/ (accessed July 18, 2015).

7. "Ethiopia's Giant Gibe Hydro Plant, Africa's Tallest Dam, Set to Produce Electricity by August," March 19, 2015, *CBS News*, http://mgafrica.com/article/2015-03-19 -ethiopias-giant-gibe-hydro-plant-set-to-produce-electricity-by-august (accessed July 18, 2015).

References

Abdullah, Sabah, and Wilner P. Jeanty. 2011. Willingness to Pay for Renewable Energy: Evidences from a Contingent Valuation Survey in Kenya. *Renewable and Sustainable Energy Reviews* 15 (6): 2974–2983.

Abdullah, Sabah, and Anil Markandya. 2009. "Rural Electrification Programmes in Kenya: Policy Conclusion from a Valuation Study." Bath Economics Research Paper 25/09.

Achen, Christopher H., and Duncan Snidal. 1989. Rational Deterrence Theory and Comparative Case Studies. *World Politics* 41 (2): 143–169.

Acker, Richard H., and Daniel M. Kammen. 1996. The Quiet (Energy) Revolution: Analysing the Dissemination of Photovoltaic Power Systems in Kenya. *Energy Policy* 24 (1): 81–111.

Adserà, Alicia, Carles Boix, and Mark Payne. 2003. Are You Being Served? Political Accountability and Quality of Government. *Journal of Law Economics and Organization* 19 (2): 445–490.

Afrane, George, and Augustine Ntiamoah. 2012. Analysis of the Life-Cycle Costs and Environmental Impacts of Cooking Fuels Used in Ghana. *Applied Energy* 98:301–306.

Africa Progress Panel. 2015. "Power, People, Planet: Seizing Africa's Energy and Climate Opportunities." Africa Progress Report 2015. Geneva.

Ailawadi, V. S., and Subhes C. Bhattacharyya. 2006. Access to Energy Services by the Poor in India: Current Situation and Need for Alternative Strategies. *Natural Resources Forum* 30 (1): 2–14.

Aklin, Michaël, and Patrick Bayer, S. P. Harish, and Johannes Urpelainen. 2014a. "Consumer Demand and Experience with Mera Gao Power in Barabanki, Uttar Pradesh." Field Survey Report. https://www.dropbox.com/s/invyeh23lbdep7z/SummerReportMGP.pdf?dl=0.

Aklin, Michaël, Patrick Bayer, S. P. Harish, and Johannes Urpelainen. 2014b. Information and Energy Policy Preferences: A Survey Experiment on Public Opinion about Electricity Pricing Reform in Rural India. *Economics of Governance* 15 (4): 305–327.

Aklin, Michaël, Patrick Bayer, S. P. Harish, and Johannes Urpelainen. 2015. Quantifying Slum Electrification in India and Explaining Local Variation. *Energy* 80 (1): 203–212.

Aklin, Michaël, Patrick Bayer, S. P. Harish, and Johannes Urpelainen. 2017. Does Basic Energy Access Generate Socioeconomic Benefits? A Field Experiment with Off-Grid Solar Power in India. *Science Advances* 3 (5): e1602153.

Aklin, Michaël, Chao-yo Cheng, Johannes Urpelainen, Karthik Ganesan, and Abhishek Jain. 2016. Factors Affecting Household Satisfaction with Electricity Supply in Rural India. *Nature Energy* 1:16170.

Alam, M. S., E. Kabir, M. M. Rahman, and M. A. K. Chowdhury. 2004. Power Sector Reform in Bangladesh: Electricity Distribution System. *Energy* 29 (11): 1773–1783.

Alkon, Meir, S. P. Harish, and Johannes Urpelainen. 2016. Household Energy Access and Expenditure in Developing Countries: Evidence from India, 1987–2010. *Energy for Sustainable Development* 35:25–34.

Alstone, Peter, Dimitry Gershenson, and Daniel M. Kammen. 2015. Decentralized Energy Systems for Clean Electricity Access. *Nature Climate Change* 5 (4): 305–314.

Amin, Narima, and Richard Langendoen. 2012. Grameen Shakti: A Renewable Energy Social Business Model for Global Replication. In *Proceedings of the IEEE Global Humanitarian Technology Conference*. Piscataway, NJ: IEEE.

Andadari, Roos Kities, Peter Mulder, and Piet Rietveld. 2014. Energy Poverty Reduction by Fuel Switching: Impact Evaluation of the LPG Conversion Program in Indonesia. *Energy Policy* 66:436–449.

Andrews, Matt, Lant Pritchett, and Michael Woolcock. 2017. *Building State Capability: Evidence, Analysis, Action*. New York: Oxford University Press.

Anozie, Ambrose N., Adekunle R. Bakare, Jacob A. Sonibare, and Timothy O. Oyebisi. 2007. Evaluation of Cooking Energy Cost, Efficiency, Impact on Air Pollution and Policy in Nigeria. *Energy* 32 (7): 1283–1290.

Apple, J., R. Vicente, A. Yarberry, N. Lohse, E. Mills, A. Jacobson, and D. Poppendieck. 2010. Characterization of Particulate Matter Size Distributions and Indoor Concentrations from Kerosene and Diesel Lamps. *Indoor Air* 20 (5): 399–411.

Arnold, J. E. Michael, Gunnar Köhlin, and Reidar Persson. 2006. Woodfuels, Livelihoods, and Policy Interventions: Changing Perspectives. *World Development* 34 (3): 596–611.

Asaduzzaman, M., F. Douglas Barnes, and Shahidur R. Khandker. 2011. "Restoring Balance: Bangladesh's Rural Energy Realities." World Bank Working Paper 181.

Asian Development Bank. 2015. *Fossil Fuel Subsidies in Indonesia: Trends, Impacts, and Reforms*. Manila, Philippines: Asian Development Bank.

Asian Institute of Technology. 2004. "Institutional Reforms and Their Impact in Rural Electrification: South and Southeast Asia." Sub Regional Technical Report, Version 10. Pathumthani, Thailand.

Baccini, Leonardo, and Johannes Urpelainen. 2014. *Cutting the Gordian Knot of Economic Reform: When and How International Institutions Help*. New York: Oxford University Press.

Bacon, R. W., and J. Besant-Jones. 2001. Global Electric Power Reform, Privatization, and Liberalization of the Electric Power Industry in Developing Countries. *Annual Review of Energy and the Environment* 26:331–359.

Badiani, Reena, Katrina K. Jessoe, and Suzanne Plant. 2012. Development and the Environment: The Implications of Agricultural Electricity Subsidies in India. *Journal of Environment and Development* 21 (2): 244–262.

Baidya, Kedar N. 1986. The Firewood Crisis in India: A Major Socio-cultural Problem of Rural Communities. *International Journal of Environmental Studies* 26 (4): 279–294.

Bailey, Frederick George. 1960. *Tribe, Caste, and Nation: A Study of Political Activity and Political Change in Highland Orissa*. Manchester: Manchester University Press.

Bailis, Rob, Amanda Cowan, Victor Berrueta, and Omar Masera. 2009. Arresting the Killer in the Kitchen: The Promises and Pitfalls of Commercializing Improved Cookstoves. *World Development* 37 (10): 1694–1705.

Balachandra, Patil. 2010. "The Status of Rural Energy Access in India: A Synthesis." Energy Technology Innovation Policy Discussion Paper (2010–09). Belfer Center for Science and International Affairs, Harvard Kennedy School, Harvard University.

Balachandra, Patil. 2011a. Dynamics of Rural Energy Access in India: An Assessment. *Energy* 36 (9): 5556–5567.

Balachandra, Patil. 2011b. Modern Energy Access to All in Rural India: An Integrated Implementation Strategy. *Energy Policy* 39:7803–7814.

Bambawale, Malavika Jain, Anthony L. D'Agostino, and Benjamin K. Sovacool. 2011. Realizing Rural Electrification in Southeast Asia: Lessons from Laos. *Energy for Sustainable Development* 15 (1): 41–48.

Banerjee, Abhijit, and Esther Duflo. 2014. "Under the Thumb of History? Political Institutions and the Scope for Action." NBER Working Paper 19848.

Banerjee, Abhijit, and Rohini Somanathan. 2007. The Political Economy of Public Goods: Some Evidence from India. *Journal of Development Economics* 82 (2): 287–314.

Banerjee, Sudeshna Ghosh, Douglas Barnes, Bipul Singh, Kristy Mayer, and Hussain Samad. 2015. *Power for All: Electricity Access Challenge in India*. Washington, DC: World Bank.

Bansal, Mohit, R. P. Saini, and D. K. Khatod. 2013. Development of Cooking Sector in Rural Areas in India: A Review. *Renewable and Sustainable Energy Reviews* 17 (1): 44–53.

Bardhan, Pranab. 2002. Decentralization of Governance and Development. *Journal of Economic Perspectives* 16 (4): 185–205.

Bardhan, Pranab, and Dilip Mookherjee. 2006a. Decentralisation and Accountability in Infrastructure Delivery in Developing Countries. *Economic Journal (Oxford)* 116 (508): 101–127.

Bardhan, Pranab, and Dilip Mookherjee. 2006b. Pro-Poor Targeting and Accountability of Local Governments in West Bengal. *Journal of Development Economics* 79 (2): 303–327.

Bardhan, Pranab, Sandip Mitra, Dilip Mookherjee, and Abhirup Sarkar. 2009. Local Democracy and Clientelism: Implications for Political Stability in Rural West Bengal. *Economic and Political Weekly* 44 (9): 46–58.

Barnes, Brendon, Angela Mathee, Elizabeth Thomas, and Nigel Bruce. 2009. Household Energy, Indoor Air Pollution and Child Respiratory Health in South Africa. *Journal of Energy in Southern Africa* 20 (1): 4–13.

Barnes, Douglas F., ed. 2007. *The Challenge of Rural Electrification: Strategies for Developing Countries*. Washington, DC: RFF Press.

Barnes, Douglas F. 2011. Effective Solutions for Rural Electrification in Developing Countries: Lessons from Successful Programs. *Current Opinion in Environmental Sustainability* 3 (4): 260–264.

Barnes, Douglas F. 2014. *Electric Power for Rural Growth: How Electricity Affects Life in Developing Countries*. 2nd ed. Washington, DC: Energy for Development.

Barnes, Douglas F., and Gerald Foley. 2004. *Rural Electrification in the Developing World: A Summary of Lessons from Successful Programs*. Joint UNDP/World Bank Energy Sector Management Assistance Programme. Washington, DC: ESMAP.

Barnes, Douglas F., Shahidur R. Khandker, and Hussain A. Samad. 2011. Energy Poverty in Rural Bangladesh. *Energy Policy* 39 (2): 894–904.

Barnes, Douglas F., Keith Openshaw, Kirk R. Smith, and Robert Van der Plas. 1994. "What Makes People Cook with Improved Biomass Stoves?" World Bank Technical Paper 242.

Barnes, Douglas F., Robert Van der Plas, and Willem Floor. 1997. Tackling the Rural Energy Problem in Developing Countries. *Finance and Development* 34 (2): 11–15.

Basalla, George. 1988. *The Evolution of Technology*. New York: Cambridge University Press.

Bates, Robert H. 1981. *Markets and States in Africa: The Political Basis of Agricultural Policies*. Berkeley: University of California Press.

Batley, Richard, and Claire Mcloughlin. 2015. The Politics of Public Services: A Service Characteristics Approach. *World Development* 74:275–285.

Baviskar, Baburao Shravan. 1980. *The Politics of Development: Sugar Co-operatives in Rural Maharashtra*. Delhi: Oxford University Press.

Bawakyillenuo, Simon. 2009. Policy and Institutional Failures: Photovoltaic Solar Household System (PV/SHS) Dissemination in Ghana. *Energy and Environment* 20 (6): 927–947.

Bayer, Patrick, and Johannes Urpelainen. 2016. It's All about Political Incentives: Democracy and the Renewable Feed-In Tariff. *Journal of Politics* 78 (2): 603–619.

Bazilian, Morgan, Smita Nakhooda, and Thijs Van de Graaf. 2014. Energy Governance and Poverty. *Energy Research and Social Science* 1 (1): 217–225.

Bazilian, Morgan, Patrick Nussbaumer, Christine Eibs-Singer, Abeeku Brew-Hammond, Vijay Modi, Benjamin Sovacool, Venkata Ramana, and Peri-Khan Aqrawi. 2012. Improving Access to Modern Energy Services: Insights from Case Studies. *Electricity Journal* 25 (1): 93–114.

Bekker, Bernard, Anton Eberhard, Charles T. Gaunt, and Andrew Marquard. 2008. South Africa's Rapid Electrification Programme: Policy, Institutional, Planning, Financing and Technical Innovations. *Energy Policy* 36 (8): 3125–3137.

Bernard, Tanguy. 2010. Impact Analysis of Rural Electrification Projects in Sub-Saharan Africa. *World Bank Research Observer* 27 (1): 33–51.

Bertocci, Peter J. 1982. Bangladesh in the Early 1980s: Praetorian Politics in an Intermediate Regime. *Asian Survey* 22 (10): 988–1008.

Besant-Jones, John E. 2006. "Reforming Power Markets in Developing Countries: What Have We Learned?" World Bank Discussion Paper 19.

Besley, Timothy, Rohini Pande, Lupin Rahman, and Vijayendra Rao. 2004. The Politics of Public Good Provision: Evidence from Indian Local Governments. *Journal of the European Economic Association* 2 (2–3): 416–426.

Besley, Timothy, Rohini Pande, and Vijayendra Rao. 2012. Just Rewards? Local Politics and Public Resource Allocation in South India. *World Bank Economic Review* 26 (2): 191–216.

Bezemer, Dirk, and Derek Headey. 2008. Agriculture, Development, and Urban Bias. *World Development* 36 (8): 1342–1364.

Bhattacharyya, Subhes C. 2006. Energy Access Problem of the Poor in India: Is Rural Electrification a Remedy? *Energy Policy* 34 (18): 3387–3397.

Bhattacharyya, Subhes C. 2013. Electrification Experiences from Sub-Saharan Africa. In *Rural Electrification through Decentralised Off-grid Systems in Developing Countries*, edited by Subhes C. Bhattacharyya, 131–156. London: Springer.

Bhattacharyya, Subhes C., and Sanusi Ohiare. 2012. The Chinese Electricity Access Model for Rural Electrification: Approach, Experience and Lessons for Others. *Energy Policy* 49:676–687.

Bhattacharya, Saugata, and Urjit R. Patel. 2007. The Power Sector in India: An Inquiry into the Efficacy of the Reform Process. *India Policy Forum* 4 (1): 211–283.

Bhattacharyya, Subhes C., and Leena Srivastava. 2009. Emerging Regulatory Challenges Facing the Indian Rural Electrification Programme. *Energy Policy* 37 (1): 68–79.

Bhutto, Abdul Waheed, and Sadia Karim. 2007. Energy Poverty Alleviation in Pakistan through Use of Indigenous Energy Resources. *Energy for Sustainable Development* 11 (1): 58–67.

Binswanger, Hans P., and Klaus Deininger. 1997. Explaining Agricultural and Agrarian Policies in Developing Countries. *Journal of Economic Literature* 35 (4): 1958–2005.

Binswanger, Hans P., and Robert F. Townsend. 2000. The Growth Performance of Agriculture in Subsaharan Africa. *American Journal of Agricultural Economics* 82 (5): 1075–1086.

Birol, Fatih. 2007. Energy Economics: A Place for Energy Poverty in the Agenda? *Energy Journal* 28 (3): 1–6.

Bonjour, Sophie, Heather Adair-Rohani, Jennyfer Wolf, Niger G. Bruce, Sumi Mehta, Annette Prüss-Ustün, Maureen Lahiff, Eva A. Rehfuess, Vinod Mishra, and Kirk R. Smith. 2013. Solid Fuel Use for Household Cooking: Country and Regional Estimates for 1980–2010. *Environmental Health Perspectives* 121 (7): 784–790.

Brass, Jennifer N., Sanya Carley, Lauren M. MacLean, and Elizabeth Baldwin. 2012. Power for Development: A Review of Distributed Generation Projects in the Developing World. *Annual Review of Environment and Resources* 37:107–136.

Brew-Hammond, Abeeku. 2010. Energy Access in Africa: Challenges Ahead. *Energy Policy* 38:2291–2301.

Briscoe, John, and R. P. S. Malik. 2006. *India's Water Economy: Bracing for a Turbulent Future*. New York: Oxford University Press.

British Petroleum. 2013. *BP Statistical Review of World Energy June 2013*. London: BP.

Brown, David S., and Ahmed Mushfiq Mobarak. 2008. The Transforming Power of Democracy: Regime Type and the Distribution of Electricity. *American Political Science Review* 103 (2): 193–213.

BTI. 2014. "Our High-Energy Planet: A Climate Pragmatism Project." Breakthrough Institute Report.

Budya, Hanung, and Muhammad Yasir Arofat. 2011. Providing Cleaner Energy Access in Indonesia through the Megaproject of Kerosene Conversion to LPG. *Energy Policy* 39 (12): 7575–7586.

Bueno de Mesquita, Bruce, Alastair Smith, Randolph M. Siverson, and James D. Morrow. 2003. *The Logic of Political Survival*. Cambridge, MA: MIT Press.

Burlig, Fiona, and Louis Preonas. 2016. "Out of the Darkness and into the Light? Development Effects of Rural Electrification in India." Energy Institute at Haas Working Paper 268.

Cabraal, Anil R., Douglas F. Barnes, and Sachin G. Agarwal. 2005. Productive Uses of Energy for Rural Development. *Annual Review of Environment and Resources* 30:117–144.

Cameron, Colin, Shonali Pachauri, Narasimha D. Rao, David McCollum, Joeri Rogelj, and Keywan Riahi. 2016. Policy Trade-Offs between Climate Mitigation and Clean Cook-Stove Access in South Asia. *Nature Energy* 1: 15010.

Carmody, Rachel N., and Richard W. Wrangham. 2009. The Energetic Significance of Cooking. *Journal of Human Evolution* 57 (4): 379–391.

Central Ground Water Board. 2012. *Ground Water Year Book: India, 2011–12*. New Delhi: Ministry of Water Resources, Government of India.

Chakravarty, Shoibal, and Massimo Tavoni. 2013. Energy Poverty Alleviation and Climate Change Mitigation: Is There a Trade Off? *Energy Economics* 40:S67–S73.

Chatterjee, Elizabeth. 2012. Dissipated Energy: Indian Electric Power and the Politics of Blame. *Contemporary South Asia* 20 (1): 91–103.

Chattopadhyay, Pradip. 2004. Cross-Subsidy in Electricity Tariffs: Evidence from India. *Energy Policy* 32 (5): 673–684.

Chaurey, Akanksha, and Tara Chandra Kandpal. 2010. Assessment and Evaluation of PV Based Decentralized Rural Electrification: An Overview. *Renewable and Sustainable Energy Reviews* 14 (8): 2266–2278.

Chaurey, Akanksha, Malini Ranganathan, and Parimita Mohanty. 2004. Electricity Access for Geographically Disadvantaged Rural Communities: Technology and Policy Insights. *Energy Policy* 32 (15): 1693–1705.

Checkel, Jeffrey T. 2005. "It's the Process Stupid! Process Tracing in the Study of European and International Politics." Arena Working Paper 26.

Cheibub, José Antonio. 1998. Political Regimes and the Extractive Capacity of Governments: Taxation in Democracies and Dictatorships. World Politics 50 (3): 349–376.

Cheng, Chao-yo, and Johannes Urpelainen. 2014. Fuel Stacking in India: Changes in the Cooking and Lighting Mix, 1987–2010. Energy 76: 306–317.

Chow, Jeffrey, Raymond J. Kopp, and Paul R. Portney. 2003. Energy Resources and Global Development. Science 302 (5650): 1528–1531.

Clark, Gregory, and David Jacks. 2007. Coal and the Industrial Revolution, 1700–1869. European Review of Economic History 2 (1): 39–72.

Coelho, Suani T., and José Goldemberg. 2013. Energy Access: Lessons Learned in Brazil and Perspectives for Replication in Other Developing Countries. Energy Policy 61:1088–1096.

Cohen, Linda R., and Roger G. Noll. 1991. The Technology Pork Barrel. Washington, DC: Brookings Institution Press.

Congleton, Roger D. 1992. Political Institutions and Pollution Control. Review of Economics and Statistics 74 (3): 412–421.

Contreras, Zaida. 2006. "Modèle d'Électrification Rurale Pour Localités de Moins de 500 Habitants au Sénégal." GTZ Study.

Cook, Cynthia C., Tyrrell Duncan, Somchai Jitsuchon, Anil Sharma, and Guobao Wu. 2005. Assessing the Impact of Transport and Energy Infrastructure on Poverty Reduction. Manila, Phillipines: Asian Development Bank.

Corbridge, Stuart. 2010. The Political Economy of Development in India since Independence. In Routledge Handbook of South Asian Politics: India, Pakistan, Bangladesh, Sri Lanka, and Nepal, edited by Paul Brass, 305–320. London: Routledge.

Correa, Paulo, Carlos Pereira, Bernardo Mueller, and Marcus Melo. 2006. "Regulatory Governance in Infrastructure Industries: Assessment and Measurement of Brazilian Regulators." World Bank PPIAF Trends and Policy Options No. 3.

Craine, Stewart, Evan Mills, and Justin Guay. 2014. Clean Energy Services for All: Financing Universal Electrification. San Francisco: Sierra Club.

Crousillat, Enrique, Richard Hamilton, and Antmann Pedro. 2010. "Addressing the Electricity Access Gap." World Bank Energy Sector Strategy Background Paper.

Crutzen, Paul J. 2006. The "Anthropocene." In Earth System Science in the Anthropocene, edited by Eckart Ehlers and Thomas Krafft, 13–18. New York: Springer.

Cullet, Philippe. 2012. "Groundwater Regulation in Uttar Pradesh: Beyond the 2010 Bill." IELRC Policy Paper.

Das, Anjana, and Jyoti Parikh. 2000. Making Maharashtra State Electricity Board Commercially Viable. *Economic and Political Weekly* 35 (14): 1201–1208.

Datye, Sunil. 1987. Marathwada Sub-Regionalism: A Factor in Maharashtra Politics. *Indian Journal of Political Science* 48 (4): 512–525.

De Oliveira, Adilson, and Tara Laan. 2010. *Lessons Learned from Brazil's Experience with Fossil-Fuel Subsidies and Their Reform.* Geneva: International Institute for Sustainable Development.

Deichmann, Uwe, Craig Meisner, Siobhan Murray, and David Wheeler. 2011. The Economics of Renewable Energy Expansion in Rural Sub-Saharan Africa. *Energy Policy* 39 (1): 215–227.

Desai, Sonalde, Amaresh Dubey, B. L. Joshi, Mitali Sen, Abusaleh Shariff, and Reeve Vanneman. 2007. "India Human Development Survey (IHDS)." University of Maryland and National Council of Applied Economic Research.

Desai, Sonalde, Amaresh Dubey, and Reeve Vanneman. 2012. *India Human Development Survey-II (IHDS II).* College Park: University of Maryland and National Council of Applied Economic Research.

Devas, Nick, and Ursula Grant. 2003. Local Government Decision-Making—Citizen Participation and Local Accountability: Some Evidence from Kenya and Uganda. *Public Administration and Development* 23 (4): 307–316.

Dholakia, Ravindra H. 2002. "Economic Reforms and Development Strategy in Gujarat." Working Paper. http://www.iimahd.ernet.in/assets/snippets/workingpaperpdf/2002–12–02RavindraHDholakia.pdf.

Differ Group. 2012. "The Indonesian Electricity System: A Brief Overview." http://www.differgroup.com/Portals/53/images/Indonesia_overall_FINAL.pdf.

Dinkelman, Taryn. 2011. The Effects of Rural Electrification on Employment: New Evidence from South Africa. *American Economic Review* 101 (7): 3078–3108.

DME. 2001. "National Electrification Programme (NEP) 1994–1999: Summary Evaluation Report." Department of Minerals and Energy, Republic of South Africa, Pretoria.

Doner, Richard F. 2009. *The Politics of Uneven Development: Thailand's Economic Growth in Comparative Perspective.* New York: Cambridge University Press.

Dossani, Rafiq. 2004. Reorganization of the Power Distribution Sector in India. *Energy Policy* 32 (11): 1277–1289.

Dubash, Navroz K. 2003. Revisiting Electricity Reform: The Case for a Sustainable Development Approach. *Utilities Policy* 11 (3): 143–154.

Dubash, Navroz K., and Sudhir Chella Rajan. 2001. Power Politics: Process of Power Sector Reform in India. *Economic and Political Weekly* 36:3367–3390.

Dubash, Navroz K., and D. Narasimha Rao. 2008. Regulatory Practice and Politics: Lessons from Independent Regulation in Indian Electricity. *Utilities Policy* 16 (4): 321–331.

Duncan, Ian. 1999. Dalits and Politics in Rural North India: The Bahujan Samaj Party In Uttar Pradesh. *Journal of Peasant Studies* 27 (1): 35–60.

Eberhard, Anton. 2004. "The Political Economy of Power Sector Reform in South Africa." Stanford University, Program on Energy and Sustainable Development, Working Paper 6.

Economist Intelligence Unit. 2014. *Powering Up: Perspectives on Indonesia's Energy Future*. London: Economist.

Edjekumhene, Ishmael, Martin Bawa Amadu, and Abeeku Brew-Hammond. 2001. "Power Sector Reform in Ghana: The Untold Story." Kumasi Institute of Technology and Environment, January.

EIA. 2015. "Country Report for Nigeria." http://www.eia.gov/countries/cab.cfm ?fips=ni.

El-Katiri, Laura, and Bassam Fattouh. 2011. *Energy Poverty in the Arab World: The Case of Yemen*. Oxford: Oxford Institute for Energy Studies.

Ellegard, Anders. 1996. Cooking Fuel Smoke and Respiratory Symptoms among Women in Low-Income Areas in Maputo. *Environmental Health Perspectives* 104 (9): 980–985.

Elojärvi, Maria, Aditya Poudyal, Mizanur Rahman, and Jukka V. Paatero. 2012. Review on Rural Energy Policy: Nepal, Ghana, Bangladesh, and Zambia. In *Proceedings of the Fourth International Conference on Sustainable Energy and Environment*. https://www.researchgate.net/profile/Md_Mizanur_Rahman2/publication/ 253954513_Review_on_Rural_Energy_Policy_Nepal_Ghana_Bangladesh_and _Zambia/links/00b7d51fb5f4b302d0000000/Review-on-Rural-Energy-Policy-Nepal -Ghana-Bangladesh-and-Zambia.pdf.

Epstein, Mollie B., Michael N. Bates, Narendra K. Arora, Kalpana Balakrishnan, Darby W. Jack, and Kirk R. Smith. 2013. Household Fuels, Low Birth Weight, and Neonatal Death in India: The Separate Impacts of Biomass, Kerosene, and Coal. *International Journal of Hygiene and Environmental Health* 216 (5): 523–532.

Erdman, Howard L. 2007. *The Swatantra Party and Indian Conservatism*. Cambridge: Cambridge University Press.

Erdogdu, Erkan. 2011. What Happened to Efficiency in Electricity Industry after Reforms? *Energy Policy* 39 (10): 6551–6560.

Fall, Abdoulaye, Sécou Sarr, Touria Dafrallah, and Abdou Ndour. 2008. Modern Energy Access in Peri-Urban Areas of West Africa: The Case of Dakar, Senegal. *Energy for Sustainable Development* 12 (4): 22–37.

Fall, Alioune, and Njeri Wamukonya. 2003. Power Sector Reform in Senegal. In *Electricity Reform: Social and Environmental Challenges*, edited by Njeri Wamukonya, 193–199. Roskilde: UNEP.

Fearon, James D. 2011. Self-Enforcing Democracy. *Quarterly Journal of Economics* 126 (4): 1661–1708.

Ferejohn, John A. 1986. Incumbent Performance and Electoral Control. *Public Choice* 50 (1): 5–25.

Foell, Wesley, Shonali Pachauri, Daniel Spreng, and Hisham Zerriffi. 2011. Household Cooking Fuels and Technologies in Developing Economies. *Energy Policy* 39 (12): 7487–7496.

Foley, Gerald. 1992. Rural Electrification in the Developing World. *Energy Policy* 20 (2): 145–152.

Forcano, Ricardo. 2003. "Removal of Barriers to the Use of Renewable Energy Sources for Rural Electrification in Chile." MSc thesis, MIT.

Foster, Vivien. 2000. "Measuring the Impact of Energy Reform: Practical Options." World Bank Private Sector and Infrastructure Network Paper.

Fouquet, Roger. 2008. *Heat Power and Light: Revolutions in Energy Services*. Cheltenham: Edward Elgar.

Fouquet, Roger. 2010. The Slow Search for Solutions: Lessons from Historical Energy Transitions by Sector and Service. *Energy Policy* 38 (11): 6586–6596.

Fox, Jonathan A. 2015. Social Accountability: What Does the Evidence Really Say? *World Development* 72:346–361.

Franda, Marcus. 1981. Ziaur Rahman and Bangladeshi Nationalism. *Economic and Political Weekly* 16 (10/12): 357–380.

Furukawa, Chishio. 2014. Do Solar Lamps Help Children Study? Contrary Evidence from a Pilot Study in Uganda. *Journal of Development Studies* 50 (2): 319–341.

GACC. 2012. "Vietnam Market Assessment: Intervention Options." Report by Accenture Development Partnerships for the Global Alliance for Clean Cookstoves.

Gales, Ben, Astrid Kander, Paolo Malanima, and Mar Rubio. 2007. North versus South: Energy Transition and Energy Intensity in Europe over 200 Years. *European Review of Economic History* 11 (2): 219–253.

Galvan, Dennis. 2001. Political Turnover and Social Change in Senegal. *Journal of Democracy* 12 (3): 51–62.

Gangopadhyay, Shubhashis, Bharat Ramaswamia, and Wilima Wadhwa. 2005. Reducing Subsidies on Household Fuels in India: How Will It Affect the Poor? *Energy Policy* 33 (18): 2326–2336.

Gaunt, Charles T. 2005. Meeting Electrification's Social Objectives in South Africa, and Implications for Developing Countries. *Energy Policy* 33 (10): 1309–1317.

Geddes, Barbara. 1999. What Do We Know about Democratization after Twenty Years? *Annual Review of Political Science* 2 (1): 115–144.

Geller, Howard, Roberto Schaeffer, Alexandre Szklo, and Mauricio Tolmasquim. 2004. Policies for Advancing Energy Efficiency and Renewable Energy Use in Brazil. *Energy Policy* 32 (12): 1437–1450.

Gencer, Defne, Peter Meier, Richard Spencer, and Hung Tien Van. 2011. *State and People, Central and Local, Working Together: The Vietnam Rural Electrification Experience*. Washington, DC: World Bank.

George, Alexander L. 1979. Case Studies and Theory Development: The Method of Structured, Focused Comparison. In *Diplomacy: New Approaches in History, Theory, and Policy*, edited by Paul Gordon Lauren, 43–68. New York: Free Press.

Gerring, John. 2004. What Is a Case Study and What Is It Good For? *American Political Science Review* 98 (2): 341–354.

Ghana Ministry of Energy. 2010. "SHEP: Ghana's Self-Help Electrification Programme." Climate Parliament: Climate Change & Energy Access for the Poor.

Ghana Statistical Service. 2014. "Ghana Living Standards Survey Round 6." Main Report.

Global Alliance for Clean Cookstoves. 2013. "Kenya Country Action Plan." http://cleancookstoves.org/resources_files/kenya-country-action-plan.pdf.

GNESD. 2014. "Energy Poverty in Developing Countries' Urban Poor Communities: Assessments and Recommendations." Case Study.

Goldemberg, José, Emilio Lèbre La Rovere, and Suani Teixeira Coelho. 2004. Expanding Access to Electricity in Brazil. *Energy for Sustainable Development* 8 (4): 86–94.

Goldemberg, José, Suani Teixeira Coelho, and Patricia Guardabassi. 2008. The Sustainability of Ethanol Production from Sugarcane. *Energy Policy* 36 (6): 2086–2097.

Goldemberg, José, Thomas B. Johansson, Amulya K. N. Reddy, and Robert H. Williams. 2004. A Global Clean Cooking Fuel Initiative. *Energy for Sustainable Development* 8 (3): 5–12.

Golumbeanu, Raluca, and Douglas Barnes. 2013. "Connection Charges and Electricity Access in Sub-Saharan Africa." World Bank Policy Research Working Paper 6511.

Gómez, Maria F., and Semida Silveira. 2010. Rural Electrification of the Brazilian Amazon: Achievements and Lessons. *Energy Policy* 38 (10): 6251–6260.

Government of India. 2011a. "2011 Census, Population Density in Indian States." http://www.census2011.co.in/density.php.

Government of India. 2011b. "2011 Census Report, Houselisting and Housing Census Data Highlights." http://www.censusindia.gov.in/2011census/hlo/hlo_highlights .html.

Government of India. 2011c. "Source of Lighting: 2001–2011." http://www .censusindia.gov.in/2011census/hlo/Data_sheet/Source%20of%20Lighting.pdf.

Government of India. 2011d. "States Census 2011." http://www.census2011.co.in/ states.php.

Government of India. 2013. "Press Note on Poverty Estimates, 2011–12." http:// planningcommission.nic.in/news/pre_pov2307.pdf.

Government of India. 2017. "Growth of Electricity Sector in India from 1947–2017." http://www.cea.nic.in/reports/others/planning/pdm/growth_2017.pdf.

Grant, Ruth W., and Robert O. Keohane. 2005. Accountability and Abuses of Power in World Politics. *American Political Science Review* 99 (1): 29–43.

Gratwick, Katharine Nawaal, and Anton Eberhard. 2008. Demise of the Standard Model for Power Sector Reform and the Emergence of Hybrid Power Markets. *Energy Policy* 36 (10): 3948–3960.

Green, Daniel. 1995. Ghana's "Adjusted" Democracy. *Review of African Political Economy* 22 (66): 577–585.

Greenstone, Michael. 2014. "Energy, Growth and Development." Evidence Paper, International Growth Centre.

Grimm, Michael, Anicet Munyehirwe, Jörg Peters, and Maximiliane Sievert. 2016. A First Step Up the Energy Ladder? Low Cost Solar Kits and Household's Welfare in Rural Rwanda. *World Bank Economic Review* 31 (3): 631–649.

Grindle, Merilee S. 2007. *Going Local: Decentralization, Democratization, and the Promise of Good Governance.* Princeton, NJ: Princeton University Press.

Grubb, Michael. 2006. "Climate Change Impacts, Energy, and Development." Paper presented at the Annual Bank Conference on Development Economics, Tokyo, May 30.

Gunaratne, Lalith. 2002. "Rural Energy Services: Best Practices." Report for USAID-SARI/Energy Program.

Gurtoo, Anjula, and Rahul Pandey. 2001. Power Sector in Uttar Pradesh: Past Problems and Initial Phase of Reforms. *Economic and Political Weekly* 36 (31): 2943–2953.

Gyimah-Boadi, Emmanuel. 1994. Ghana's Uncertain Political Opening. *Journal of Democracy* 5 (2): 75–86.

Haanyika, Charles M. 2006. Rural Electrification Policy and Institutional Linkages. *Energy Policy* 34 (17): 2977–2993.

Haanyika, Charles M. 2008. Rural Electrification in Zambia: A Policy and Institutional Analysis. *Energy Policy* 36 (3): 1044–1058.

Hagopian, Frances, Carlos Gervasoni, and Juan Andres Moraes. 2009. From Patronage to Program: The Emergence of Party-Oriented Legislators in Brazil. *Comparative Political Studies* 42 (3): 360–391.

Hall, David. 2007. "Energy Privatisation and Reform in East Africa." PSIRU Reports.

Hanna, Rema, Esther Duflo, and Michael Greenstone. 2016. Up in Smoke: The Influence of Household Behavior on the Long-Run Impact of Improved Cooking Stoves. *American Economic Journal. Economic Policy* 8 (1): 80–114.

Hansen, Christopher J., and John Bower. 2003. "Political Economy of Electricity Reform: A Case Study in Gujarat, India." Oxford Institute for Energy Studies Working Paper.

Hausman, William J., Peter Hertner, and Mira Wilkins. 2011. *Global Electrification: Multinational Enterprise and International Finance in the History of Light and Power, 1878–2007*. New York: Cambridge University Press.

Hausman, William J., John L. Neufeld, and Till Schreiber. 2014. Multilateral and Bilateral Aid Policies and Trends in the Allocation of Electrification Aid, 1970–2001. *Utilities Policy* 29:54–62.

Heltberg, Rasmus. 2004. Fuel Switching: Evidence from Eight Developing Countries. *Energy Economics* 26 (5): 869–887.

Hiremath, Rahul B., Bimlesh Kumar, Palit Balachandra, Nijavalli H. Ravindranath, and Basavanahalli N. Raghunandan. 2009. Decentralised Renewable Energy: Scope, Relevance and Applications in the Indian Context. *Energy for Sustainable Development* 13 (1): 4–10.

Hodler, Roland, and Paul A. Raschky. 2014. Regional Favoritism. *Quarterly Journal of Economics* 129 (2): 995–1033.

Hofman, Bert, and Kai Kaiser. 2006. Decentralization, Democratic Transition, and Local Governance in Indonesia. In *Decentralization and Local Governance in Developing Countries: A Comparative Perspective*, edited by Pranab Bardhan and Dilip Mookherjee, 81–124. Cambridge, MA: MIT Press.

Hopwood, Bill, Mary Mellor, and Geoff O'Brien. 2005. Sustainable Development: Mapping Different Approaches. *Sustainable Development* 13 (1): 38–52.

Hornsby, Charles. 2013. *Kenya: A History since Independence*. New York: I. B. Tauris.

Hossain, Abdul K., and Ossama Badr. 2007. Prospects of Renewable Energy Utilisation for Electricity Generation in Bangladesh. *Renewable and Sustainable Energy Reviews* 11 (8): 1617–1649.

Hughes, Thomas P. 1993. *Networks of Power: Electrification in Western Society, 1880–1930*. Baltimore: Johns Hopkins University Press.

Hummels, David. 2007. Transportation Costs and International Trade in the Second Era of Globalization. *Journal of Economic Perspectives* 21 (3): 131–154.

IBRD. 2013. *Toward Universal Access to Clean Cooking*. Indonesia: International Bank for Reconstruction and Development.

IDA. 1972. "Report and Recommendation of the President to the Executive Directors on a Proposed Credit to the Republic of Indonesia for a Second Electricity Distribution Project." International Development Association.

IEA. 2002. *World Energy Outlook*. Paris: International Energy Agency.

IEA. 2009. *Chile Energy Policy Review*. Paris: International Energy Agency.

IEA. 2010a. *Comparative Study on Rural Electrification Policies in Emerging Economies: Keys to Successful Policies*. Paris: International Energy Agency.

IEA. 2010b. *World Energy Outlook*. Paris: International Energy Agency.

IEA. 2011. *World Energy Outlook*. Paris: International Energy Agency.

IEA. 2012. Oil and Gas Security: Emergency Response of IEA Countries; Chile. Paris: International Energy Agency.

IEA. 2014a. *Africa Energy Outlook*. Paris: International Energy Agency.

IEA. 2014b. "World Energy Outlook." Electricity Access Database (OECD/IEA).

IEA. 2017. "Energy Access Outlook 2017: From Poverty to Prosperity." World Energy Outlook Special Report.

Ikejemba, Eugene, and Peter Schuur. 2016. Locating Solar and Wind Parks in South-Eastern Nigeria for Maximum Population Coverage: A Multi-Step Approach. *Renewable Energy* 89:449–462.

Ikeme, James, and Obas John Ebohon. 2005. Nigeria's Electric Power Sector Reform: What Should Form the Key Objectives? *Energy Policy* 33 (9): 1213–1221.

India Ministry of New and Renewable Energy. 2014. *Implementation of Unnat Chulha Abhiyan (UCA) Programme*. Note.

IPCC. 2007. *Contribution of Working Group III to the Fourth Assessment Report of the Intergovernmental Panel on Climate Change*. Cambridge: Cambridge University Press.

IPCC. 2013. "Climatic Change 2013: The Physical Science Basis. Contribution of Working Group I to the Fifth Assessment Report of the Intergovernmental Panel on Climate Change." Summary for Policymakers.

IRENA. 2012. "Senegal: Renewables Readiness Assessment." Report.

Islam, Akhtarul K. M. Sadrul, Mazharul Islam, and Tazmilur Rahman. 2006. Effective Renewable Energy Activities in Bangladesh. *Renewable Energy* 31 (5): 677–688.

Islamic Development Bank. 2013. "From Darkness to Light: Rural Electricity in Morocco." IsDB Success Story Series.

Jadresic, Alejandro. 2000. A Case Study of Subsidizing Rural Electrification in Chile. In *Energy and Development Report: Energy Services for the World's Poor*, edited by Penelope J. Brooke and Suzanne Smith, 76–82. Washington, DC: World Bank.

Jannuzzi, Gilberto M., and Godfrey A. Sanga. 2004. LPG Subsidies in Brazil: An Estimate. *Energy for Sustainable Development* 8 (3): 127–129.

Jannuzzi, Gilberto M., and José Goldemberg. 2014. Modern Energy Services to Low-Income Households in Brazil. In *Energy Poverty: Global Challenges and Local Solutions*, edited by Antoine Halff, Benjamin K. Sovacool, and Jon Rozhon, 257–270. Oxford: Oxford University Press.

Jaoul, Nicolas. 2006. Learning the Use of Symbolic Means Dalits, Ambedkar Statues and the State in Uttar Pradesh. *Contributions to Indian Sociology* 40 (2): 175–207.

Jewitt, Sarah, and Sujatha Raman. 2017. Energy Poverty, Institutional Reform and Challenges of Sustainable Development: The Case of India. *Progress in Development Studies* 17 (2): 173–185.

Joon, Vinod, Avinash K. Chandra, and Madhulekha Bhattacharya. 2009. Household Energy Consumption Pattern and Socio-Cultural Dimensions Associated with It: A Case Study of Rural Haryana, India. *Biomass and Bioenergy* 33 (11): 1509–1512.

Joseph, Kelli L. 2010. The Politics of Power: Electricity Reform in India. *Energy Policy* 38 (1): 503–511.

Kale, Sunila S. 2014a. *Electrifying India: Regional Political Economies of Development*. Stanford: Stanford University Press.

Kale, Sunila S. 2014b. Structures of Power: Electrification in Colonial India. *Comparative Studies of South Asia, Africa and the Middle East* 34 (3): 454–475.

Kale, Sunila S., and Nimah Mazaheri. 2014. Natural Resources, Development Strategies, and Lower Caste Empowerment in India's Mineral Belt: Bihar and Odisha during the 1990s. *Studies in Comparative International Development* 49 (3): 343–369.

Kandpal, Jai B., Ramesh C. Maheshwari, and Tara Chandra Kandpal. 1995. Indoor Air Pollution from Domestic Cookstoves Using Coal, Kerosene and LPG. *Energy Conversion and Management* 36 (11): 1067–1072.

Kaneko, Shinji, Satoru Komatsu, and Partha Pratim Ghosh. 2012. Rural Electrification in Bangladesh: Implications for Climate Change Mitigation. In *Climate Change Mitigation and International Development*, edited by Ryo Fujikura and Tomoyo Toyota, 202–226. Oxford: Earthscan.

Kapika, Joseph, and Anthon Eberhard. 2013. *Power-Sector Reform and Regulation in Africa: Lessons from Kenya, Tanzania, Uganda, Zambia, Namibia and Ghana*. Cape Town: HSRC Press.

Karekezi, Stephen. 2002. Poverty and Energy in Africa: A Brief Review. *Energy Policy* 30 (11–12): 915–919.

Karekezi, Stephen, and Donella Mutiso. 2000. Power Sector Reform: A Kenyan Case Study. In *Power Sector Reform in Sub-Saharan Africa*, edited by John K. Turkson, 83–120. New York: Springer.

Kaundinya, Deepak Paramashivan, Palit Balachandra, and Nijavalli H. Ravindranath. 2009. Grid-Connected versus Stand-Alone Energy Systems for Decentralized Power: A Review of Literature. *Renewable and Sustainable Energy Reviews* 138:2041–2050.

Keefer, Philip. 2007. Clientelism, Credibility, and the Policy Choices of Young Democracies. *American Journal of Political Science* 51 (4): 804–821.

Kemausuor, Francis, George Yaw Obeng, Abeeku Brew-Hammond, and Alfred Duker. 2011. A Review of Trends, Policies and Plans for Increasing Energy Access in Ghana. *Renewable and Sustainable Energy Reviews* 15 (9): 5143–5154.

Kemmler, Andreas. 2006. "Regional Disparities in Electrification of India: Do Geographic Factors Matter?" Center for Energy Policy and Economics Working Paper 51.

Kemmler, Andreas. 2007. Factors Influencing Household Access to Electricity in India. *Energy for Sustainable Development* 11 (4): 3–20.

Kenyan Ministry of Energy and Petroleum. 2015. National Energy and Petroleum Policy—Final Draft. June 16, 2015.

Khandker, Shahidur R., Douglas F. Barnes, and Hussain A. Samad. 2009. "Welfare Impacts of Rural Electrification: A Case Study from Bangladesh." World Bank Policy Research Working Paper 4859.

Khandker, Shahidur R., Douglas F. Barnes, and Hussain A. Samad. 2012. Are the Energy Poor Also Income Poor? Evidence from India. *Energy Policy* 47:1–12.

Khandker, Shahidur R., Douglas F. Barnes, and Hussain A. Samad. 2013. Welfare Impacts of Rural Electrification: A Panel Data Analysis from Vietnam. *Economic Development and Cultural Change* 61 (3): 659–692.

Khandker, Shahidur R., Hussain A. Samad, Rubaba Ali, and Douglas Barnes. 2012. "Who Benefits Most from Rural Electrification? Evidence in India." World Bank Policy Research Working Paper 6095.

Killick, Tony. 1989. *A Reaction Too Far: Economic Theory and the Role of the State in Developing Countries*. London: Overseas Development Institute.

Kim, Sung Eun, and Johannes Urpelainen. 2013. When and How Can Advocacy Groups Promote New Technologies? Conditions and Strategies for Effectiveness. *Journal of Public Policy* 33 (3): 259–293.

Kim, Sung Eun, and Johannes Urpelainen. 2014. Technology Competition and International Co-operation: Friends or Foes? *British Journal of Political Science* 44 (3): 545–574.

Kimmich, Christian. 2013a. Incentives for Energy-Efficient Irrigation: Empirical Evidence of Technology Adoption in Andhra Pradesh, India. *Energy for Sustainable Development* 17:261–269.

Kimmich, Christian. 2013b. Linking Action Situations: Coordination, Conflicts, and Evolution in Electricity Provision for Irrigation in Andhra Pradesh, India. *Ecological Economics* 90:150–158.

King, Gary, Robert O. Keohane, and Sidney Verba. 1994. *Designing Social Inquiry: Scientific Inference in Qualitative Research*. Princeton: Princeton University Press.

Kiplagat, Jeremiah K., Ruzhu Z. Wang, and Tingxian X. Li. 2011. Renewable Energy in Kenya: Resource Potential and Status of Exploitation. *Renewable and Sustainable Energy Reviews* 15 (6): 2960–2973.

Kirubi, Charles, Arne Jacobson, Daniel M. Kammen, and Andrew Mills. 2009. Community-Based Electric Micro-Grids Can Contribute to Rural Development: Evidence from Kenya. *World Development* 37 (7): 1208–1221.

Kitschelt, Herbert. 2000. Linkages between Citizens and Politicians in Democratic Polities. *Comparative Political Studies* 33 (6–7): 845–879.

Kohli, Atul. 1987. *The State and Poverty in India: The Politics of Reform*. New York: Cambridge University Press.

Kohli, Atul. 2009. *Democracy and Development in India: From Socialism to Pro-Business*. New York: Oxford University Press.

Kojima, Masami, Robert Bacon, and Chris Trimble. 2014. "Political Economy of Power Sector Subsidies: A Review with Reference to Sub-Saharan Africa." World Bank Working Paper 89547.

Kramon, Eric, and Daniel N. Posner. 2013. Who Benefits from Distributive Politics? How the Outcome One Studies Affects the Answer One Gets. *Perspectives on Politics* 11 (2): 461–474.

Kroth, Verena, Valentino Larcinese, and Joachim Wehner. 2016. A Better Life for All? Democratization and Electrification in Post-Apartheid South Africa. *Journal of Politics* 78 (3): 774–791.

Kudo, Yuya, Abu S. Shonchoy, and Kazushi Takahashi. 2017. Can Solar Lanterns Improve Youth Academic Performance? Experimental Evidence from Bangladesh. *World Bank Economic Review*. https://doi.org/10.1093/wber/lhw073.

Kumar, Atul, Parimita Mohanty, Debajit Palit, and Akanksha Chaurey. 2009. Approach for Standardization of Off-Grid Electrification Projects. *Renewable and Sustainable Energy Reviews* 13 (8): 1946–1956.

Kwon, Eunkyung. 2006. "Infrastructure, Growth, and Poverty Reduction in Indonesia: A Cross-Sectional Analysis." Working Paper, Asian Development Bank, Manila, Phillipines.

Laan, Tara, Christopher Beaton, and Bertille Presta. 2010. *Strategies for Reforming Fossil-Fuel Subsidies: Practical Lessons from Ghana, France and Senegal*. Geneva: International Institute for Sustainable Development.

Lake, David A., and Matthew A. Baum. 2001. The Invisible Hand of Democracy: Political Control and the Provision of Public Services. *Comparative Political Studies* 34 (6): 587–621.

Lal, Sumir. 2006. "Can Good Economics Ever Be Good Politics? Case Study of India's Power Sector." World Bank Working Paper 83.

Lalvani, Mala. 2008. Sugar Co-operatives in Maharashtra: A Political Economy Perspective. *Journal of Development Studies* 44 (10): 1474–1505.

Lam, Nicholas L., Yanju Chen, Cheryl Weyant, Chandra Venkataraman, Pankaj Sadavarte, Michael A. Johnson, Kirk R. Smith, et al. 2012. Household Light Makes Global Heat: High Black Carbon Emissions from Kerosene Wick Lamps. *Environmental Science and Technology* 46 (24): 13531–13538.

Lam, Nicholas L., Kirk R. Smith, Alison Gauthier, and Michael N. Bates. 2012. Kerosene: A Review of Household Uses and their Hazards in Low- and Middle-Income Countries. *Journal of Toxicology and Environmental Health, Part B* 15 (6): 396–432.

Lambe, Fiona, and Aaron Atteridge. 2012. "Putting the Cook before the Stove: A User-Centred Approach to Understanding Household Decision-Making." Stockholm Environmental Institute Working Paper 2012–03.

Lazarus, Michael, Souleymane Diallo, and Youba Sokona. 1994. Energy and Environment Scenarios for Senegal. *Natural Resources Forum* 18 (1): 31–47.

Leach, Gerald. 1987. Household Energy in South Asia. *Biomass* 12 (3): 155–184.

Lee, Kenneth, Eric Brewer, Carson Christiano, Francis Meyo, Edward Miguel, Matthew Podolsky, Javier Rosa, and Catherine Wolfram. 2014. "Barriers to Electrification for 'Under Grid' Households in Rural Kenya." NBER Working Paper 20327.

Lee, Kenneth, Eric Brewer, Carson Christiano, Francis Meyo, Edward Miguel, Matthew Podolsky, Javier Rosa, and Catherine Wolfram. 2016. Electrification for "Under Grid" Households in Rural Kenya. *Development Engineering* 1:26–35.

Lenz, Luciane, Anicet Munyehirwe, Jörg Peters, and Maximiliane Sievert. 2017. Does Large-Scale Infrastructure Investment Alleviate Poverty? Impacts of Rwanda's Electricity Access Roll-Out Program. *World Development* 89 (Suppl. C): 88–110.

Lerche, Jens. 2000. Dimensions of Dominance: Class and State in Uttar Pradesh. In *The Everyday State and Society in Modern India*, edited by C. J. Fuller, and Veronique Benei, 91–114. London: Hurst.

Levine, David I., and Carolyn Cotterman. 2012. "What Impedes Efficient Adoption of Products? Evidence from Randomized Variation in Sales Offers for Improved Cookstoves in Uganda." Working Paper Series, Institute for Research on Labor and Employment, University of California, Berkeley.

Levy, Brian. 2014. *Working with the Grain: Integrating Governance and Growth in Development Strategies*. New York: Oxford University Press.

Levy, Jack S. 2008. Case Studies: Types, Designs, and Logics of Inference. *Conflict Management and Peace Science* 25 (1): 1–18.

Li, Quan, and Rafael Reuveny. 2006. Democracy and Environmental Degradation. *International Studies Quarterly* 50 (4): 935–956.

Lieberman, Evan S., Daniel N. Posner, and Lily L. Tsai. 2014. Does Information Lead to More Active Citizenship? Evidence from an Education Intervention in Rural Kenya. *World Development* 60:69–83.

Lieten, G. K., and Ravi Srivastava. 1999. *Power Relations, Devolution and Development in Uttar Pradesh*. Thousand Oaks, CA: Sage.

Lim, S. S., T. Vos, A. D. Flaxman, G. Danaei, K. Shibuya, H. Adair-Rohani, M. Amann, et al. 2013. A Comparative Risk Assessment of Burden of Disease and Injury Attributable to 67 Risk Factors and Risk Factor Clusters in 21 Regions, 1990–2010: A Systematic Analysis for the Global Burden of Disease Study 2010. *Lancet* 380 (9859): 2224–2260.

Lipton, Michael. 1977. *Why Poor People Stay Poor: A Study of Urban Bias in World Development*. London: Temple Smith.

Lock, Reinier. 1996. Liberalization of India's Electric Power Sector: Evolution or Anarchy? *Electricity Journal* 9 (2): 78–86.

Lora, Eduardo Silva, and Rubenildo Viera Andrade. 2009. Biomass as Energy Source in Brazil. *Renewable and Sustainable Energy Reviews* 13 (4): 777–788.

Lucon, Oswaldo, Suani Teixeira Coelho, and José Goldemberg. 2004. LPG in Brazil: Lessons and Challenges. *Energy for Sustainable Development* 8 (3): 1241–1278.

Madureira, Nuno Luís. 2008. When the South Emulates the North: Energy Policies and Nationalism in the Twentieth Century. *Contemporary European History* 17 (1): 1–21.

Maliti, Emmanuel, and Raymond Mnenwa. 2011. "Affordability and Expenditure Patterns for Electricity and Kerosene in Urban Households in Tanzania." Research on Poverty Alleviation, Research Report 11/2.

Mansuri, Ghazala, and Vijayendra Rao. 2013. *Localizing Development: Does Participation Work?* Washington, DC: World Bank.

Mariía, Domingo Santa. 1945. Chilean State Electrification. *Journal of Land and Public Utility Economics* 21 (4): 365–370.

Marquard, Andrew. 2006. "The Origins and Development of South African Energy Policy." PhD dissertation, University of Cape Town.

Marro, Peter, and Natalie Bertsch. 2015. "Making Renewable Energy A Success in Bangladesh: Getting the Business Model Right." ADB South Asia Working Paper 41.

Martinelli, Luiz A., and Solange Filoso. 2008. Expansion of Sugarcane Ethanol Production in Brazil: Environmental and Social Challenges. *Ecological Applications* 18 (4): 885–898.

Martinot, Eric, Akanksha Chaurey, Debra Lew, José Moreira, and Njrei Wamukonya. 2002. Renewable Energy Markets in Developing Countries. *Annual Review of Energy and the Environment* 27:309–348.

Masera, Omar R., Barbara D. Saatkamp, and Daniel M. Kammen. 2000. From Linear Fuel Switching to Multiple Cooking Strategies: A Critique and Alternative to the Energy Ladder Model. *World Development* 28 (12): 2083–2103.

Mathavan, Deeptha. 2008. From Dabhol to Ratnagiri: The Electricity Act of 2003 and Reform of India's Power Sector. *Columbia Journal of Transnational Law* 47 (2): 387.

Mawhood, Rebecca. 2012. "The Senegalese Rural Electrification Action Plan: A 'Good Practice' Model for Increasing Private Sector Participation in Sub-Saharan Rural Electrification?" MSc thesis, Imperial College London.

Mawhood, Rebecca, and Robert Gross. 2014. Institutional Barriers to a "Perfect" Policy: A Case Study of the Senegalese Rural Electrification Plan. *Energy Policy* 73:480–490.

McAllistair, Joseph A., and Daniel B. Waddle. 2007. Rural Electricity Subsidies and the Private Sector in Chile. In *The Challenge of Rural Electrification: Strategies for Developing Countries*, edited by Douglas F. Barnes, 198–224. Washington, DC: Resource for the Future.

McCarthy, John D., and Mayer N. Zald. 1977. Resource Mobilization and Social Movements: A Partial Theory. *American Journal of Sociology* 82 (6): 1212–1241.

McCawley, Peter. 1970. The Price of Electricity. *Bulletin of Indonesian Economic Studies* 6 (3): 61–86.

McCawley, Peter. 1978. Rural Electrification in Indonesia: Is It Time? *Bulletin of Indonesian Economic Studies* 14 (2): 34–69.

Mestl, Heidi Elizabeth Staff, Kristin Aunan, Hans Martin Seip, Zhao Shuxiao Wang Yu, and Daisheng Zhang. 2007. Urban and Rural Exposure to Indoor Air Pollution from Domestic Biomass and Coal Burning across China. *Science of the Total Environment* 377 (1): 12–26.

Miah, Md. Danesh, Harun Al Rashid, and Man Yong Shin. 2009. Wood Fuel Use in the Traditional Cooking Stoves in the Rural Floodplain Areas of Bangladesh: A Socio-Environmental Perspective. *Biomass and Bioenergy* 33 (1): 70–78.

Mill, John Stuart. 2011. *A System of Logic, Ratiocinative and Inductive: Being a Connected View of the Principles of Evidence, and the Methods of Scientific Investigation.* Cambridge: Cambridge University Press.

Miller, Grant, and A. Mushfiq Mobarak. 2011. "Intra-household Externalities and Low Demand for a New Technology: Experimental Evidence on New Cookstoves." Working Paper, Stanford University and Yale University.

Millinger, Markus, Tina Møarlind, and Erik O. Ahlgren. 2012. Evaluation of Indian Rural Solar Electrification: A Case Study in Chhattisgarh. *Energy for Sustainable Development* 16 (4): 486–492.

Min, Brian. 2015. *Power and the Vote: Electricity and Politics in the Developing World.* New York: Cambridge University Press.

Min, Brian, and Miriam Golden. 2014. Electoral Cycles in Electricity Losses in India. *Energy Policy* 65:619–625.

Mishra, Vinod K., Robert D. Retherford, and Kirk R. Smith. 1999. Biomass Cooking Fuels and Prevalence of Tuberculosis in India. *International Journal of Infectious Diseases* 3 (3): 119–129.

MNRE. 2009. "National Biomass Cookstoves Programme." http://www.mnre.gov.in/ schemes/decentralized-systems/national-biomass-cookstoves-initiative/.

MNRE. 2013. "Physical Progress of Implementation of Remote Village Electrification Programme." Accessed March 28, 2018. https://mnre.gov.in/remote-village -electrification.

Mobarak, Ahmed Mushfiq, Puneet Dwivedi, Robert Bailis, Lynn Hildemann, and Grant Miller. 2012. Low Demand for Nontraditional Cookstove Technologies. *Proceedings of the National Academy of Sciences of the United States of America* 109 (27): 10815–10820.

Modi, Vijay. 2005. "Improving Electricity Services in Rural India." Center on Globalization and Sustainable Development Working Paper 30.

Monari, Lucio. 2002. "Power Subsidies: A Reality Check on Subsidizing Power for Irrigation in India." World Bank Note 244.

Mookherjee, Dilip. 2015. Political Decentralization. *Annual Review of Economics* 7:231–249.

Moss, Todd, and Benjamin Leo. 2014. *Maximizing Access to Energy: Estimates of Access and Generation for the Overseas Private Investment Corporation's Portfolio.* Washington, DC: Center for Global Development.

Moss, Todd, Roger A. Pielke, and Morgan Bazilian. 2014. "Balancing Energy Access and Environmental Goals in Development Finance: The Case of the OPIC Carbon Cap." Center for Global Development Policy Paper 038.

Mueller, Susanne D. 2008. The Political Economy of Kenya's Crisis. *Journal of Eastern African Studies: Journal of the British Institute in Eastern Africa* 2 (2): 185–210.

Mukherji, Aditi. 2006. Political Ecology of Groundwater: The Contrasting Case of Water-Abundant West Bengal and Water-Scarce Gujarat, India. *Hydrogeology Journal* 14:392–406.

Mukherji, Aditi, and Arijit Das. 2014. The Political Economy of Metering Agricultural Tube Wells in West Bengal, India. *Water International* 39 (5): 671–685.

Mukherji, Aditi, Tushaar Shah, and Shilp Verma. 2010. Electricity Reforms and Its Impact on Groundwater Use: Evidence from India. In *Re-thinking Water and Food Security: Fourth Botin Foundation Water Workshop,* edited by Luis Martinez-Cortina, Alberto Garrido, and Elena Lopez-Gunn, 299–306. Boca Raton, FL: CRC Press.

Munasinghe, Mohan. 1988. Rural Electrification International Experience and Policy in Indonesia. *Bulletin of Indonesian Economic Studies* 24 (2): 87–105.

Mwabu, Germano, and Erik Thorbecke. 2004. Rural Development, Growth and Poverty in Africa. *Journal of African Economies* 13:116–165.

Najam, Adil. 2005. Developing Countries and Global Environmental Governance: From Contestation to Participation to Engagement. *International Environmental Agreement: Politics, Law and Economics* 5 (3): 303–321.

Najam, Adil, Saleemul Huq, and Youba Sokona. 2003. Climate Negotiations beyond Kyoto: Developing Countries Concerns and Interests. *Climate Policy* 3 (2): 221–231.

Narayan, Badri. 2009. *Fascinating Hindutva: Saffron Politics and Dalit Mobilisation.* Thousand Oaks, CA: Sage.

Neumayer, Eric. 2002. Do Democracies Exhibit Stronger International Environmental Commitment? A Cross-Country Analysis. *Journal of Peace Research* 39 (2): 139–164.

Niez, Alexandra. 2010. "Comparative Study on Rural Electrification Policies in Emerging Economies: Keys to Successful Policies." IEA Information Paper.

Noxie Consult. 2010. "Electricity Access Progress in Ghana." Report.

NREL. 2015. Policies to Spur Energy Access: Volume 2; Case Studies of Public-Private Models to Finance Decentralized Electricity Access. Technical Report NREL/TP-7A40-64460. National Renewable Energy Laboratory, September.

NSS. 2012a. Energy Sources of Indian Households for Cooking and Lighting. NSS Report 542. National Sample Survey Office, Government of India.

NSS. 2012b. Household Consumption of Various Goods and Services in India. NSS Report 541. National Sample Survey Office, Government of India.

Nygaard, Ivan. 2010. Institutional Options for Rural Energy Access: Exploring the Concept of the Multifunctional Platform in West Africa. *Energy Policy* 38 (2): 1192–1201.

Nyoike, Patrick. 2002. Is the Kenyan Electricity Regulatory Board Autonomous? *Energy Policy* 30 (11–12): 987–997.

Nyukuri, Baras Kundu. 1997. "The Impact of Past and Potential Ethnic Conflicts on Kenyan's Stability and Development." Paper presented at the USAID Conference on Conflict Resolution in the Greater Horn of Africa, June.

Obermaier, Martin, Alexandre Szklo, Emilio L. La Rovere, and Luiz P. Rosa. 2012. An Assessment of Electricity and Income Distributional Trends Following Rural Electrification in Poor Northeast Brazil. *Energy Policy* 49:531–540.

ODI. 2016. "Beyond Coal: Scaling Up Clean Energy to Fight Global Poverty." Overseas Development Institute Position Paper.

Ogot, Bethwell A., and William R. Ochieng. 1996. *Decolonization and Independence in Kenya 1940–93.* Columbus: Ohio University Press.

Ohiare, Sanusi. 2015. Expanding Electricity Access to All in Nigeria: A Spatial Planning and Cost Analysis. *Energy, Sustainability and Society* 5 (8): 1–18.

Okoro, Ogbonnaya I., and Ed Chikuni. 2007. Power Sector Reforms in Nigeria: Opportunities and Challenges. *Journal of Energy in Southern Africa* 18 (3): 52–57.

Olken, Benjamin A. 2010. Direct Democracy and Local Public Goods: Evidence from a Field Experiment in Indonesia. *American Political Science Review* 104 (2): 243–267.

Olson, Mancur. 1993. Dictatorship, Democracy, and Development. *American Political Science Review* 87 (3): 567–576.

Olukoju, Ayodeji. 2004. "Never Expect Power Always": Electricity Consumers' Response to Monopoly, Corruption and Inefficient Services in Nigeria. *African Affairs* 103 (1): 51–71.

Ondraczek, Janosh. 2011. "The Sun Rises in the East (of Africa): A Comparison of the Development and Status of the Solar Energy Markets in Kenya and Tanzania." Working Paper FNU-195.

Onyeji, Ijeoma, Morgan Bazilian, and Patrick Nussbaumer. 2012. Contextualizing Electricity Access in Sub-Saharan Africa. *Energy for Sustainable Development* 16 (4): 520–527.

Opam, Michael. 1995. "Institution Building in the Energy Sector of Africa: A Case Study of the Ghana Power Sector Reform Programme." Paper presented at the Sixth Session of the Africa Regional Conference on Mineral and Energy Resources Development and Utilisation, Accra, Ghana, November 14–23.

Oseni, Musiliu O. 2011. An Analysis of the Power Sector Performance in Nigeria. *Renewable and Sustainable Energy Reviews* 15 (9): 4765–4774.

Oseni, Musiliu O. 2012. Households' Access to Electricity and Energy Consumption Pattern in Nigeria. *Renewable and Sustainable Energy Reviews* 16 (1): 990–995.

Oyedepo, Sunday Olayinka. 2012. On Energy for Sustainable Development in Nigeria. *Renewable and Sustainable Energy Reviews* 16 (5): 2583–2598.

Pachauri, Shonali. 2014. Household Electricity Access a Trivial Contributor to CO_2 Emissions Growth in India. *Nature Climate Change* 4:1073–1076.

Pachauri, Shonali, and Leiwen Jiang. 2008. The Household Energy Transition in India and China. *Energy Policy* 36 (11): 4022–4035.

Pachauri, S., A. Mueller, A. Kemmler, and D. Spreng. 2004. On Measuring Energy Poverty in Indian Households. *World Development* 32 (12): 2083–2104.

Pachauri, Shonali, and Daniel Spreng. 2004. Energy Use and Energy Access in Relation to Poverty. *Economic and Political Weekly* 39 (3): 17–23.

Pachauri, Shonali, and Daniel Spreng. 2011. Measuring and Monitoring Energy Poverty. *Energy Policy* 39:7497–7504.

Pai, Sudha. 2002. *Dalit Assertion and the Unfinished Democratic Revolution: The Bahujan Samaj Party in Uttar Pradesh.* Thousand Oaks, CA: Sage.

Palit, Debajit. 2013. Solar Energy Programs for Rural Electrification: Experiences and Lessons from South Asia. *Energy for Sustainable Development* 17 (3): 270–279.

Palit, Debajit, Subhes C. Bhattacharyya, and Akanksha Chaurey. 2014. Indian Approaches to Energy Access. In *Energy Poverty: Global Challenges and Local Solutions*, edited by Antoine Halff, Benjamin K. Sovacool and Jon Rozhon, 237–256. New York: Oxford University Press.

Palit, Debajit, and Akanksha Chaurey. 2011. Off-Grid Rural Electrification Experiences from South Asia: Status and Best Practices. *Energy for Sustainable Development* 15 (3): 266–276.

Palit, Debajit, Gopal K. Sarangi, and P. R. Krithika. 2014. Energising Rural India Using Distributed Generation: The Case of Solar Mini-Grids in Chhattisgarh State, India. In *Mini-Grids for Rural Electrification of Developing Countries*, edited by Subhes C. Bhattacharyya and Debajit Palit, 313–342. London: Springer.

Palit, Debajit, Benjamin K. Sovacool, Christopher Cooper, David Zoppo, Jay Eidsness, Meredith Crafton, Katie Johnson, and Shannon Clarke. 2013. The Trials and Tribulations of the Village Energy Security Programme (VESP) in India. *Energy Policy* 57:407–417.

Palshikar, Suhas, and Rajeshwari Deshpande. 1999. Electoral Competition and Structures of Domination in Maharashtra. *Economic and Political Weekly* 34 (34/35): 2409–2422.

Pan, Jiahua, Wuyuan Peng, Meng Li, Xiangyang Wu, Lishuang Wan, Hisham Zerriffi, David Victor, Becca Elias, and Chi Zhang. 2006. "Rural Electrification in China 1950–2004: Historical Processes and Key Driving Forces." Stanford University, Program on Energy and Sustainable Development Working Paper 60.

Pandey, M. R., K. R. Smith, J. S. M. Boleij, and E. M. Wafula. 1989. Indoor Air Pollution in Developing Countries and Acute Respiratory Infection in Children. *Lancet* 333 (8635): 427–429.

Parikh, Jyoti, Kirk Smith, and Vijay Laxmi. 1999. Indoor Air Pollution: A Reflection on Gender Bias. *Economic and Political Weekly* 34 (9): 539–544.

Patel, V. J. 1985. Rational Approach towards Fuelwood Crisis in Rural India. *Economic and Political Weekly* 20 (32): 1366–1368.

Paul, Samuel. 1992. Accountability in Public Services: Exit, Voice and Control. *World Development* 20 (7): 1047–1060.

Peng, Wuyuan, and Jiahua Pan. 2006. Rural Electrification in China: History and Institution. *China and World Economy* 14 (1): 71–84.

Pereira, Marcio Giannini, Aureélio Vasconcelos Freitas Marcos, and Neilton Fidelis da Silva. 2010. Rural Electrification and Energy Poverty: Empirical Evidences from Brazil. *Renewable and Sustainable Energy Reviews* 14 (4): 1229–1240.

Pokharel, Shaligram. 2001. Hydropower for Energy in Nepal. *Mountain Research and Development* 21 (1): 4–9.

Pollitt, Michael. 2004. Electricity Reform in Chile: Lessons for Developing Countries. Working Paper 2004-016. Cambridge, MA: MIT Center for Energy and Environmental Policy Research.

Posel, Deborah. 1991. *The Making of Apartheid, 1948–1961: Conflict and Compromise.* New York: Oxford University Press.

Poten & Partners. 2003. *The Story of LPG.* London: Poten & Partners.

Powell, Stephen, and Mary Starks. 2000. Does Reform of Energy Sector Networks Improve Access for the Poor? Note no. 209. The World Bank Group, Private Sector and Infrastructure Network.

Practical Action. 2016. *Poor People's Energy Outlook 2016.* Bourton on Dunsmore, UK: Practical Action Publishing.

Practical Action. 2017. *Poor People's Energy Outlook 2017.* Bourton on Dunsmore, UK: Practical Action Publishing.

Prahalad, Coimbatore Krishnarao. 2006. *The Fortune at the Bottom of the Pyramid.* Upper Saddle River:, NJ Pearson Education.

Prasad, Gisela. 2008. Energy Sector Reform, Energy Transitions and the Poor in Africa. *Energy Policy* 36 (8): 2806–2811.

Prins, Gwyn, Isabel Galiana, Christopher Green, Reiner Grundmann, Mike Hulme, Atte Korhola, Frank Laird, et al. 2010. The Hartwell Paper: A New Direction for Climate Policy after the Crash of 2009. University of Oxford and London School of Economics and Political Science.

Pritchett, Lant. 2005. "A Lecture on the Political Economy of Targeted Safety Nets." World Bank Social Protection Discussion Paper Series 0501.

Przeworski, Adam. 1991. *Democracy and the Market: Political and Economic Reform in Eastern Europe and Latin America.* New York: Cambridge University Press.

Przeworski, Adam, and Henry Teune. 1970. *The Logic of Comparative Social Inquiry.* Malabar, FL: Krieger.

Psacharopoulos, George, Samuel Morley, Ariel Fiszbein, Haeduck Lee, and William C. Wood. 1995. Poverty and Income Inequality in Latin America during the 1980s. *Review of Income and Wealth* 41 (3): 245–264.

PT Pertamina, and World LP Gas Association. 2015. "Kerosene to LP Gas Conversion Programme in Indonesia: A Case Study of Domestic Energy."

Purohita, Pallav, and Axel Michaelowa. 2007. CDM Potential of Bagasse Cogeneration in India. *Energy Policy* 35 (10): 4779–4798.

Quaye-Foli, Emmanuel A. 2002. "Liquified Petroleum Gas (LPG) Promotion: The Ghana Experience." Presentation at UNDP/World Bank Energy and Poverty Workshop, Addis Ababa.

Rahman, Md. Mizanur, Jukka V. Paatero, Aditya Poudyal, and Risto Lahdelma. 2013. Driving and Hindering Factors for Rural Electrification in Developing Countries: Lessons from Bangladesh. *Energy Policy* 61:840–851.

Rajan, Thillai A. 2000. Power Sector Reform in Orissa: An Ex-Post Analysis of the Causal Factors. *Energy Policy* 28 (10): 657–669.

Ranganathan, V. 1996. Electricity Privatization Revisited: A Commentary on the Case for New Initiatives in India. *Energy Policy* 24 (9): 821–825.

Rao, Narasimha D. 2012. Kerosene Subsidies in India: When Energy Policy Fails as Social Policy. *Energy for Sustainable Development* 16 (1): 35–43.

Rao, Yenda Srinivasa. 2010. Electricity, Politics, and Regional Economic Imbalance in Madras Presidency. *Economic and Political Weekly* 45 (23): 59–66.

Ravetz, Alison. 1968. The Victorian Coal Kitchen and Its Reformers. *Victorian Studies* 11 (4): 435–460.

Ray, Amal. 1974. Sub-Regional Politics and Elections in Orissa. *Economic and Political Weekly* 9 (49): 2032–2036.

Reddy, B. Sudhakara, Palit Balachandra, and Hippu Salk Kristle Nathan. 2009. Universalization of Access to Modern Energy Services in India Households: Economic and Policy Analysis. *Energy Policy* 37 (11): 4645–4657.

Rehman, Ibrahim Hafeezur, Abhishek Kar, Manjushree Banerjee, Preeth Kumar, Martand Shardul, Jeevan Mohanty, and Ijaz Hossain. 2012. Understanding the Political Economy and Key Drivers of Energy Access in Addressing National Energy Access Priorities and Policies. *Energy Policy* 47 (S1): 27–37.

Rehman, Ibrahim Hafeezur, and Preeti Malhotra, eds. 2004. *Fire without Smoke: Learning from the National Programme on Improved Chulhas.* New Delhi: Energy and Resources Institute.

Rehman, Ibrahim Hafeezur, Preeti Malhotra, Ram Chandra Pal, and Phool Badan Singh. 2005. Availability of Kerosene to Rural Households: A Case Study from India. *Energy Policy* 33 (17): 2165–2174.

Repetto, Robert. 1994. *The "Second India" Revisited: Population, Poverty, and Environmental Stress over Two Decades*. Washington, DC: World Resources Institute.

Revelle, Roger. 1976. Energy Use in Rural India. *Science* 192 (4243): 969–975.

Rodrigue, Jean-Claude, and Claude Comtois. 2013. *The Geography of Transport Systems*. New York: Routledge.

Rud, Juan Pablo. 2012. Electricity Provision and Industrial Development: Evidence from India. *Journal of Development Economics* 97 (2): 352–367.

Rufín, Carlos, U. Srinivasa Rangan, and Rajesh Kumar. 2003. The Changing Role of the State in the Electricity Industry in Brazil, China, and India. *American Journal of Economics and Sociology* 62 (4): 649–675.

Rural Electrification Board. 2009. "Rural Electrification: Bangladesh Experience." African Electrification Initiative Practitioner Workshop.

Sadath, Anver C., and Rajesh H. Acharya. 2017. Assessing the Extent and Intensity of Energy Poverty Using Multidimensional Energy Poverty Index: Empirical Evidence from Households in India. *Energy Policy* 102 (Suppl. C): 540–550.

Sagar, Ambuj D. 2005. Alleviating Energy Poverty for the World's Poor. *Energy Policy* 33 (11): 1367–1372.

Sala-i-Martin, Xavier, and Arvind Subramanian. 2013. Addressing the Natural Resource Curse: An Illustration from Nigeria. *Journal of African Economies* 22 (4): 570–615.

Samad, Hussain A., and Shahidur R. Khandker. M. Asaduzzaman, and Mohammad Yunus. 2013. "The Benefits of Solar Home Systems: An Analysis from Bangladesh." World Bank Policy Research Working Paper 6724.

Samanta, B. B., and A. K. Sundaram. 1983. "Socioeconomic Impact of Rural Electrification in India." Discussion Paper D-730, Resources for the Future, Washington DC.

Samuels, David, and Richard Snyder. 2001. The Value of a Vote: Malapportionment in Comparative Perspective. *British Journal of Political Science* 31 (4): 651–671.

Sanoh, Aly, Lily Parshall, Ousmane Fall Sarr, Susan Kum, and Vijay Modi. 2012. Local and National Electricity Planning in Senegal: Scenarios and Policies. *Energy for Sustainable Development* 16 (1): 13–25.

Schei, Morten A., Jens O. Hessen, Kirk R. Smith, Nigel Bruce, John McCracken, and Victorina Lopez. 2004. Childhood Asthma and Indoor Woodsmoke from

Cooking in Guatemala. *Journal of Exposure Analysis and Environmental Epidemiology* 14:S110–S117.

Schlag, Nicolai, and Fiona Zuzarte. 2008. "Market Barriers to Clean Cooking Fuels in Sub-Saharan Africa: A Review of the Literature." SEI Working Paper.

Schurr, Sam H., and Bruce Carlton Netschert. 1960. *Energy in the American Economy, 1850–1975.* Baltimore, MD: Johns Hopkins University Press.

SE4ALL. 2011. "Sustainable Energy for All." Report.

SE4ALL. 2017a. "Global Tracking Framework: Progress Towards Sustainable Energy." Report.

SE4ALL. 2017b. "Why Wait? Seizing the Energy Access Dividend." Report.

Seawright, Jason, and John Gerring. 2008. Case Selection Techniques in Case Study Research: A Menu of Qualitative and Quantitative Options. *Political Research Quarterly* 61 (2): 294–308.

Sehjpal, Ritika, Aditya Ramji, Anmol Soni, and Atul Kumar. 2014. Going Beyond Incomes: Dimensions of Cooking Energy Transitions in Rural India. *Energy* 68:470–477.

Sekhon, Jasjeet S. 2004. Quality Meets Quantity: Case Studies, Conditional Probability, and Counterfactuals. *Perspectives on Politics* 2 (2): 281–293.

Sekhri, Sheetal. 2011. Public Provision and Protection of Natural Resources: Groundwater Irrigation in Rural India. *American Economic Journal. Applied Economics* 3 (4): 29–55.

Sell, Susan K. 1996. North-South Environmental Bargaining: Ozone, Climate Change, and Biodiversity. *Global Governance* 2 (1): 97–118.

Sen, Amartya K. 1999. *Development as Freedom.* New York: Oxford University Press.

Sesan, Temilade. 2012. Navigating the Limitations of Energy Poverty: Lessons from the Promotion of Improved Cooking Technologies in Kenya. *Energy Policy* 47:202–210.

Shah, Tushaar, Christopher Scott, Avinash Kishore, and Abhishek Sharma. 2004. "Energy-Irrigation Nexus in South Asia: Improving Groundwater Conservation and Power Sector Viability." International Water Management Institute, Research Report 70, Revised Second Edition.

Shah, Tushaar, and Shilip Verma. 2008. Co-Management of Electricity and Groundwater: An Assessment of Gujarat's Jyotirgram Scheme. *Economic and Political Weekly* 43 (7): 59–66.

Shamir, Ronen. 2013. *Current Flow: The Electrification of Palestine.* Stanford: Stanford University Press.

Sharma, D., Parameswara, P. S. Chandramohanan Nair, and R. Balasubramanian. 2005. Performance of Indian Power Sector During a Decade under Restructuring: A Critique. *Energy Policy* 33 (4): 563–576.

Shiu, Alice, and Pun-Lee Lam. 2004. Electricity Consumption and Economic Growth in China. *Energy Policy* 32 (1): 47–54.

Shukla, Sachin, and Sreyamsa Bairiganjan. 2011. *The Base of Pyramid Distribution Challenge: Evaluating Alternate Distribution Models of Energy Products for Rural Base of Pyramid in India*. Chennai, India: Centre for Development Finance.

Singer, Hans W. 1953. The Brazilian Salte Plan: An Historical Case Study of the Role of Internal Borrowing in Economic Development. *Economic Development and Cultural Change* 1 (5): 341–349.

Singh, Anoop. 2006. Power Sector Reform in India: Current Issues and Prospects. *Energy Policy* 34 (16): 2480–2490.

Singh, Nirvikar, and T. Srinivasan. 2004. "Indian Federalism, Economic Reform and Globalization." Working Paper. http://econpapers.repec.org/paper/wpawuwppe/0412007.htm.

Sinton, Jonathan E., Kirk R. Smith, John W. Peabody, Yaping Liu, Xiliang Zhang, Rufus Edwards, and Gan Quan. 2004. An Assessment of Programs to Promote Improved Household Stoves in China. *Energy for Sustainable Development* 8 (3): 33–52.

Skolnikoff, Eugene B. 1993. *The Elusive Transformation: Science, Technology, and the Evolution of International Politics*. Princeton: Princeton University Press.

Slough, Tara, Johannes Urpelainen, and Joonseok Yang. 2015. "Light for All? Evaluating Brazil's Rural Electrification Program, 2000-2010." Columbia University Working Paper.

Smil, Vaclav. 2010. *Energy Transitions: History, Requirements, Prospects*. Santa Barbara, CA: Praeger.

Smith, Kirk R. 2000. National Burden of Disease in India from Indoor Air Pollution. *Proceedings of the National Academy of Sciences of the United States of America* 97 (24): 13286–13293.

Smith, Kirk R, and Ambuj Sagar. 2014. Making the Clean Available: Escaping India's Chulha Trap. *Energy Policy* 75:410–414.

Smith, Thomas B. 2004. Electricity Theft: A Comparative Analysis. *Energy Policy* 32 (18): 2067–2076.

Snyder, Richard. 2001. Scaling Down: The Subnational Comparative Method. *Studies in Comparative International Development* 36 (1): 93–110.

Soete, Luc. 1985. International Diffusion of Technology, Industrial Development and Technological Leapfrogging. *World Development* 13 (3): 409–422.

Sokona, Youba, and Alassane Deme. 2003. LPG Subsidies in Senegal. In *Energy Subsidies: Lessons Learned in Assessing Their Impact and Designing Policy Reforms*, edited by UNEP, 113–122. Geneva: UNEP.

Solow, Robert M. 1993. Sustainability: An Economist's Perspective. In *Economics of the Environment*, edited by R. Dorfman and N. S. Dorfman, 179–187. New York: Norton.

Soo, Min Lee, Yeon-Su Kim, Wanggi Jaung, Sitti Latifah, Mansur Afifi, and Larry A. Fisher. 2015. Forests, Fuelwood and Livelihood: Energy Transition Patterns in Eastern Indonesia. *Energy Policy* 85:61–70.

Sovacool, Benjamin K. 2012a. Deploying Off-Grid Technology to Eradicate Energy Poverty. *Science* 338:47–48.

Sovacool, Benjamin K. 2012b. The Political Economy of Energy Poverty: A Review of Key Challenges. *Energy for Sustainable Development* 16 (3): 272–282.

Sovacool, Benjamin K., and Ira Martina Drupady. 2011. Summoning Earth and Fire: The Energy Development Implications of Grameen Shakti (GS) in Bangladesh. *Energy* 36 (7): 4445–4459.

Sperling, Daniel. 1987. Brazil, Ethanol and the Process of System Change. *Energy* 12 (1): 11–23.

Srivastava, Leena, and I. H. Rehman. 2006. Energy for Sustainable Development in India: Linkages and Strategic Direction. *Energy Policy* 34 (5): 643–654.

SSA. 2013. "Energy 2002–2012: In-Depth Analysis of the General Household Survey Data." Statistics South Africa, Pretoria.

Stein, Arthur A. 1980. The Politics of Linkage. *World Politics* 33 (1): 62–81.

Stern, Nicholas. 2006. *The Economics of Climate Change: The Stern Review*. Cambridge: Cambridge University Press.

Stokes, Susan C. 2005. Perverse Accountability: A Formal Model of Machine Politics with Evidence from Argentina. *American Political Science Review* 99 (3): 315–325.

Taniguchi, Mariko, and Shinji Kaneko. 2009. Operational Performance of the Bangladesh Rural Electrification Program and Its Determinants with a Focus on Political Interference. *Energy Policy* 37 (6): 2433–2439.

Teitelbaum, Emmanuel. 2011. *Mobilizing Restraint: Democracy and Industrial Conflict in Post-Reform South Asia*. Ithaca, NY: Cornell University Press.

Tenenbaum, Bernard, Chris Greacen, Tilak Siyambalapitiya, and James Knuckles. 2014. *From the Bottom Up: How Small Power Producers and Mini-Grids Can Deliver Electrification and Renewable Energy in Africa*. Washington, DC: World Bank.

Thachil, Tariq. 2011. Embedded Mobilization: Nonstate Service Provision as Electoral Strategy in India. *World Politics* 63 (3): 434–469.

Thakur, Tripta, S. G. Deshmukh, S. C. Kaushik, and Mukul Kulshrestha. 2005. Impact Assessment of the Electricity Act 2003 on the Indian Power Sector. *Energy Policy* 33 (9): 1187–1198.

Thiam, Djiby-Racine. 2010. Renewable Decentralized in Developing Countries: Appraisal from Microgrids Project in Senegal. *Renewable Energy* 35 (8): 1615–1623.

Thomas, John W., and Merilee S. Grindle. 1990. After the Decision: Implementing Policy Reforms in Developing Countries. *World Development* 18 (8): 1163–1181.

Thompson, Griffin, and Morgan Bazilian. 2014. Democratization, Energy Poverty, and the Pursuit of Symmetry. *Global Policy* 5 (1): 127–131.

Trace, Simon. 2015. Measuring Access for Different Needs. In *International Energy and Poverty: The Emerging Contours*, edited by Lakshman Guruswamy, 160–178. London: Taylor and Francis.

Treisman, Daniel. 2007. *The Architecture of Government: Rethinking Political Decentralization.* New York: Cambridge University Press.

Tsai, Lily L. 2007. *Accountability without Democracy: Solidary Groups and Public Goods Provision in Rural China.* New York: Cambridge University Press.

Uddin, Sk Noim, and Ros Taplin. 2009. Trends in Renewable Energy Strategy Development and the Role of CDM in Bangladesh. *Energy Policy* 37 (1): 281–289.

UNDP. 2004. Liquefied Petroleum Gas (LPG) Substitution for Wood Fuel in Ghana: Opportunities and Challenges. Infolink no. 1.

UNDP. 2009. "Energy in National Decentralization Policies." A Review Focusing on Least Developed Countries and Sub-Saharan Africa.

University of Ghana. 2005. "Guide to Electric Power in Ghana." Resource Center for Energy Economic and Regulation Report.

Urpelainen, Johannes. 2014. Grid and Off-Grid Electrification: An Integrated Model with Applications to India. *Energy for Sustainable Development* 19:66–71.

Vagliasindi, Maria, and John E. Besant-Jones. 2013. *Power Market Structure: Revisiting Policy Options.* Washington, DC: World Bank, Directions in Development Energy and Mining.

Valer, L. Roberto, André Mocelin, Roberto Zilles, Edila Moura, and A. Claudeise S. Nascimento. 2014. Assessment of Socioeconomic Impacts of ASccess to Electricity in Brazilian Amazon: Case Study in Two Communities in Mamirauà Reserve. *Energy for Sustainable Development* 20:58–65.

Van Arkadie, Brian, and Raymond Mallon. 2004. *Viet Nam: A Transition Tiger?* Canberra: Asia Pacific Press, Australian National University.

van Els, Rudi Henri, João Nildo de Souza Vianna, and Antonio Cesar Pinho Brasil Jr. 2012. The Brazilian Experience of Rural Electrification in the Amazon with Decentralized Generation: The Need to Change the Paradigm from Electrification to Development. *Renewable and Sustainable Energy Reviews* 16 (3): 1450–1461.

Van't Veld, Klaas, Urvashi Narain, Shreekant Gupta, Neetu Chopra, and Supriya Singh. 2006. "India's Firewood Crisis Re-examined." Resources for the Future Discussion Paper 06-25.

Varshney, Ashutosh. 1993. Urban Bias in Perspective. In *Beyond Urban Bias*, edited by Ashutosh Varshney. London: Frank Cass.

Vatikiotis, Michael R J. 1999. *Indonesian Politics under Suharto: The Rise and Fall of the New Order*. New York: Routledge.

Venkataraman, Chandra, Ambuj D. Sagar, Gazala Habib, Nick Lam, and Kirk R. Smith. 2010. The Indian National Initiative for Advanced Biomass Cookstoves: The Benefits of Clean Combustion. *Energy for Sustainable Development* 14:63–72.

Victor, David G. 2005. *The Effects of Power Sector Reform on Energy Services for the Poor*. New York: UN, Department of Economic and Social Affairs.

Victor, David G., and Thomas C. Heller, eds. 2007. *The Political Economy of Power Sector Reform: The Experiences of Five Major Developing Countries*. New York: Cambridge University Press.

Victor, David G., Joshua C. House, and Sarah Joy. 2005. A Madisonian Approach to Climate Policy. *Science* 309: 1820–1821.

Vicziany, Marika. 2002. The BJP and the Shiv Sena: A Rocky Marriage? *South Asia* 25 (3): 41–60.

Viswanathan, Brinda, and K. S. Kavi Kumar. 2005. Cooking Fuel Use Patterns in India: 1983–2000. *Energy Policy* 33 (8): 1021–1036.

Wade, Robert. 1985. The Market for Public Office: Why the Indian State Is Not Better at Development. *World Development* 13 (4): 467–497.

Wickramasinghe, Anoja. 2011. Energy Access and Transition to Cleaner Cooking Fuels and Technologies in Sri Lanka: Issues and Policy Limitations. *Energy Policy* 39 (12): 7567–7574.

Wintrobe, Ronald. 1998. *The Political Economy of Dictatorship*. New York: Cambridge University Press.

Wittman, Donald A. 1995. *The Myth of Democratic Failure: Why Political Institutions Are Efficient*. Chicago: University of Chicago Press.

Wolfram, Catherine, Orie Shelef, and Paul Gertler. 2012. How Will Energy Demand Develop in the Developing World? *Journal of Economic Perspectives* 26 (1): 119–137.

Woolcock, Michael. 1998. Social Capital and Economic Development: Toward a Theoretical Synthesis and Policy Framework. *Theory and Society* 27 (2): 151–208.

World Bank. 1995. "Power Project: Upper Indravati Hydro." World Bank. http://www.worldbank.org/projects/P009805/power-project-upper-indravati-hydro?lang=en.

World Bank. 2000. "Energy Services for the World's Poor." World Bank Energy and Development Report.

World Bank. 2002. *India: Household Energy, Air Pollution, and Health.* Washington, DC: World Bank.

World Bank. 2005. *Electricity for All: Options for Increasing Access in Indonesia.* Jakarta: World Bank Office.

World Bank. 2008. "The Welfare Impact of Rural Electrification: A Reassessment of the Costs and Benefits." Impact Evaluation Report by Independent Evaluation Group.

Wüstenhagen, Rolf, Maarten Wolsink, and Mary Jean Bürer. 2007. Social Acceptance of Renewable Energy Innovation: An Introduction to the Concept. *Energy Policy* 35 (5): 2683–2691.

Yadoo, Annabel, and Heather Cruickshank. 2010. The Value of Cooperatives in Rural Electrification. *Energy Policy* 38 (6): 2941–2947.

Zerriffi, Hisham. 2008. From Açaí to Access: Distributed Electrification in Rural Brazil. *International Journal of Energy Sector Management* 2 (1): 90–117.

Zerriffi, Hisham. 2011. *Rural Electrification: Strategies for Distributed Generation.* New York: Springer.

Zerriffi, Hisham, and Elizabeth Wilson. 2010. Leapfrogging over Development? Promoting Rural Renewables for Climate Change Mitigation. *Energy Policy* 38 (4): 1689–1700.

Zhang, Fan. 2014. Can Solar Panels Leapfrog Power Grids? The World Bank Experience 1992–2009. *Renewable and Sustainable Energy Reviews* 38:811–820.

Zhang, Lixiao, Zhifeng Yang, Bin Chen, and Guoqian Chen. 2009. Rural Energy in China: Pattern and Policy. *Renewable Energy* 34 (12): 2813–2823.

Zhao, Jimin. 2001. "Reform of China's Energy Institutions and Policies: Historical Evolution and Current Challenges." BCSIA Discussion Paper 2001-20, Energy Technology Innovation Project, Kennedy School of Government, Harvard University.

Index

Page numbers in italic and bold refer to figures and tables, respectively.